Laser beam melting of immiscible FeMn-AgX for adapted bioresorbability

zur Erlangung des akademischen Grades
DOKTOR DER INGENIEURWISSENSCHAFTEN (Dr.-Ing.)
der Fakultät für Maschinenbau
der Universität Paderborn

genehmigte
DISSERTATION

von

Jan Tobias Krüger, M. Sc.

aus Bielefeld

Tag des Kolloquiums: 06.04.2023
Referent: Prof. Dr.-Ing. habil. Mirko Schaper
Korreferent: Prof. Dr. med. vet. Manfred Kietzmann

Bibliografische Information der Deutschen Nationalbibliothek
Die Deutsche Nationalbibliothek verzeichnet diese Publikation in der Deutschen Nationalbibliografie; detaillierte bibliografische Daten sind im Internet über dnb.dnb.de abrufbar.

Zugl.: Paderborn, Univ., Diss., 2023

© 2023 Jan Tobias Krüger
Herstellung und Verlag: BoD – Books on Demand, Norderstedt

ISBN: 9783734707308

Kurzzusammenfassung

Der Einsatz von Eisen zur Erschließung neuer Anwendungsmöglichkeiten bioresorbierbarer Implantate ist vielversprechend, jedoch ist eine Modifikation notwendig, um die erforderlichen Eigenschaften einzustellen. Zur Verbesserung der mechanischen Eigenschaften und Erhöhung der Degradationsrate kann Legieren mit Mangan erfolgen. Eine weitere Beschleunigung der Degradation kann durch Einstellen elektrochemisch edler Phasen erzielt werden, die eine anodische Auflösung der eisenbasierten Matrix forcieren. Da Silber und Eisen ineinander unlöslich sind, können Silberphasen in einer unveränderten Eisen-Mangan-Matrix existieren. Silber ist biokompatibel und wirkt antibakteriell. Wird anstelle von reinem Silber eine abbaubare Silberlegierung verwendet, können die Silberphasen nach der Eisen-Mangan-Matrix degradieren.

Eine Möglichkeit schmelzmetallurgisch nicht ineinander lösliche Metalle zu verarbeiten, ist das pulverbettbasierte selektive Laserstrahlschmelzen. Das kleine Schmelzbad, der starke Schmelzfluss und die schnelle Erstarrung ermöglichen den Einschluss von Silber in der eisenbasierten Matrix. Um eine gezielte Anpassung von Morphologie, Verteilung und chemischer Zusammensetzung der Silberphasen zu ermöglichen, wurde ein Modell zur Wechselwirkung der unlöslichen Komponenten im Schmelzbad und der Entstehung der Silberphasen während des Laserstrahlschmelzens entwickelt. Die prinzipielle Effektivität der anodischen Auflösung durch Silberphasen ist bestätigt worden. Allerdings resultiert daraus keine gesteigerte Degradationsrate, da Ablagerungen mit Sperrwirkung entstehen.

Abstract

The application of Iron to open up new applications for bioresorbable implants is promising. However, modification of Iron is necessary to adjust the required properties. Alloying with Manganese improves the mechanical performance and increases the degradation rate. Further acceleration of degradation can be adjusting electrochemically noble phases to force the anodic dissolution of the Iron-based matrix. Since Silver and Iron are insoluble in each other, Silver phases can exist in an unmodified Iron-Manganese matrix. Silver is biocompatible and provides an antibacterial effect. If a degradable Silver alloy is used, the Silver phases can degrade after the Iron-Manganese matrix.

One way to process metals that are not soluble in each other by melting metallurgy is powder-bed-based selective laser beam melting. The small melt pool, strong melt flow, and rapid solidification enable the inclusion of Silver in the Iron-Manganese matrix. To enable selective tailoring of the morphology, distribution, and chemical composition of the Silver phases, a model was developed for the interaction of the insoluble components in the melt pool and the formation of the Silver phases during laser beam melting. The principle effectiveness of anodic dissolution by Silver phases has been confirmed. However, this does not result in an increased degradation rate, since deposits with a blocking effect are formed.

Danksagung

Die vorliegende Dissertation entstand während meiner Tätigkeit als wissenschaftlicher Mitarbeiter am Lehrstuhl für Werkstoffkunde (LWK) der Universität Paderborn. Für die Finanzierung des Projektes „Entwicklung und Charakterisierung bioresorbierbarer FeMnAg-Werkstoffe für den SLM-Prozess", auf dem diese Arbeit basiert, danke ich der Deutschen Forschungsgemeinschaft (DFG).

Herrn Prof. Dr.-Ing. habil. Mirko Schaper danke ich für die Möglichkeit zur Promotion am LWK. Zudem möchte ich Mirko für den Freiraum zur Umsetzung eigener Ideen und die Unterstützung in der Zeit der Promotion danken.

Bei Herrn Prof. Dr. med. vet. Manfred Kietzmann bedanke ich mich für die Mitwirkung in der Prüfungskommission und die erkenntnisreichen, wissenschaftlichen Diskussionen bei den Projekttreffen. Darüber hinaus danke ich dem gesamten Team der Stiftung Tierärztliche Hochschule Hannover für die gute Zusammenarbeit.

Meine Wertschätzung für die Übernahme des Vorsitzes der Prüfungskommission gebührt Frau Prof. Dr.-Ing. Iryna Mozgova.

Mein ganz besonderer Dank gilt Herrn Dr.-Ing. Kay-Peter Hoyer für die Unterstützung während der gesamten Zeit am LWK, insbesondere für das Korrekturlesen, die Mitwirkung in der Prüfungskommission und die Beantwortung der unzählbaren Fragen, die ich während der gemeinsamen Zeit immer wieder gestellt habe. Ohne Peter wäre ich nicht auf die Idee gekommen überhaupt zu promovieren. Aber nicht nur fachlich konnte ich mich stets auf ihn verlassen. Durch die guten und schwierigen Phasen wurde ich fürsorglich begleitet. Vielen Dank für das stets offene Ohr und die großartige gemeinsame Zeit in unserem humorvollen Büro.

Dem außergewöhnlichen Team am LWK mit tollen Kollegen und Studierenden danke ich für die gute und gewinnbringende Zusammenarbeit. Ohne Unterstützung, sei es durch die Diskussion wissenschaftlicher Inhalte, einen Beitrag durch studentische Arbeiten oder die Unterstützung bei der Durchführung von Versuchen, hätte ich die Herausforderung nicht bewältigen können. Auch für das gute Miteinander während der Corona-Zeit danke ich dem Team am LWK. Meine Wertschätzung gilt ganz besonders Anja, Barbara, Denise, Eva, Florian, Kristina, Mario, Martin, Maxwell, Sabrina, Steffen und Thomas. Zudem hat mir der freundschaftliche Austausch über die Höhen und Tiefen der Promotion geholfen. Dafür und für die fachliche Unterstützung danke ich meinen Kollegen und Freunden Anatolii, Caro, Lea, Malte und Moritz.

Meiner Familie gilt meine Wertschätzung für die emotionale Unterstützung während meiner Promotion. Auf meine Eltern, meine Großmutter, meinen Bruder und meinen Onkel konnte ich mich stets verlassen. Ich danke für euer Verständnis, dass die Zeit für die Familie hin und wieder zu kurz gekommen ist.

Bei meinen Freunden Dima, Fabian, Jenni, Maike und Norman bedanke ich mich für offene Ohren und die gemeinsame Zeit, in der ich auf andere Gedanken gekommen bin und neue Energie für die Fortsetzung dieser Arbeit sammeln konnte.

Paderborn, 12.04.2023 *Jan Tobias Krüger*

Publications

J.T. Krüger, K.P. Hoyer, V. Filor, S. Pramanik, M. Kietzmann, J. Meißner, M. Schaper, Novel AgCa and AgCaLa alloys for Fe-based bioresorbable implants with adapted degradation, Journal of Alloys and Compounds 871 (2021) 159544. https://doi.org/10.1016/j.jallcom.2021.159544

J.T. Krüger, K.P. Hoyer, M. Schaper, Bioresorbable AgCe and AgCeLa alloys for adapted Fe-based implants, Materials Letters 306 (2022) 130890. https://doi.org/10.1016/j.matlet.2021.130890

J. Huang, A.G. Orive, J.T. Krüger, K.P. Hoyer, A. Keller, G. Grundmeier, Influence of proteins on the corrosion of a conventional and selective laser beam melted FeMn alloy in physiological electrolytes, Corrosion Science 200 (2022) 110186. https://doi.org/10.1016/j.corsci.2022.110186

R. Kupfer, D. Köhler, D. Römisch, S. Wituschek, L. Ewenz, J. Kalich, D. Weiß, B. Sadeghian, M. Busch, J. Krüger, M. Neuser, O. Grydin., M. Böhnke, C.-R. Bielak, J. Troschitz, Clinching of Aluminum Materials – Methods for the Continuous Characterization of Process, Microstructure and Properties, Journal of Advanced Joining Processes 5 (2022) 100108. https://doi.org/10.1016/j.jajp.2022.100108

J.T. Krüger, K.P. Hoyer, F. Hengsbach, M. Schaper, Formation of insoluble silver-phases in an iron-manganese matrix for bioresorbable implants using varying laser-beam-melting-strategies, Journal of Materials Research and Technology 19C (2022) 2369-2387. http://dx.doi.org/10.1016/j.jmrt.2022.06.006

M. Neuser, F. Kappe, J. Ostermeier, J.T. Krüger, M. Bobbert, G. Meschut, M. Schaper, O. Grydin, Mechanical Properties and Joinability of AlSi9 Alloy Manufactured by Twin-Roll Casting, Advanced Engineering Materials, (2022) 2200874. https://doi.org/10.1002/adem.202200874

J.T. Krüger, Adjustment of AgCaLa Phases in a FeMn Matrix via LBM for Implants with Adapted Degradation, Crystals 12(8) (2022) 1146. https://doi.org/10.3390/cryst12081146

J.T. Krüger, K.-P. Hoyer, A. Andreiev, M. Schaper, C. Zinn, Modification of Iron with Degradable Silver Phases Processed via Laser Beam Melting for Implants with Adapted Degradation Rate, Advanced Engineering Materials, (2022) 2201008. https://doi.org/10.1002/adem.202201008

J.T. Krüger, K.-P. Hoyer, J. Huang, V. Filor, R.H. Mateus-Vargas, H. Oltmanns, J. Meißner, G. Grundmeier, M. Schaper, FeMn with Phases of a Degradable Ag Alloy for Residue-Free and Adapted Bioresorbability, Journal of Functional Biomaterials 13(4) (2022) 185. https://doi.org/10.3390/jfb13040185

Presentations

Krüger, J., K.-P. Hoyer, M. Schaper, Eisenbasierte Werkstoffe für die additive Fertigung von Implantaten mit angepasster Degradation. DGM - Fachtagung Werkstoffe und Additive Fertigung. 14.05.2020. online / Potsdam

Krüger, J., K.-P. Hoyer, M. Schaper, Degradation additiv verarbeiteter Eisen-Silber-Werkstoffe für resorbierbare Implantate. DVM - Workshop - Zuverlässigkeit von Implantaten und Biostrukturen. 26.11.2020. online / Berlin

Popp, E., Krüger, J., K.-P. Hoyer, M. Schaper, Additiv verarbeitete Eisen-Silber-Werkstoffe mit angepasster Mikrostruktur für resorbierbare Implantate. WTK-Kolloquium. 24.03.2021. online / Chemnitz

Krüger, J., K.-P. Hoyer, M. Schaper, Degradation of additively processed iron-silver materials for resorbable implants. DVM - Workshop - Zuverlässigkeit von Implantaten und Biostrukturen. 05.05.2021. online

Krüger, J., M. Dreyer, K.-P. Hoyer, M. Schaper, Processing of immiscible iron-silver-materials via laser beam melting. TMS 2022 Annual Meeting & Exhibition. 03.2022. online / Anaheim, California, USA

Table of contents

1 Introduction

The research in materials for implantation aims to improve the benefits for patients. An increasing number of diseases and injuries can be treated with implants, but therefore adapted implants are required. [1-5] Bioresorbable implants for temporal applications degrade and are resorbed by the human organism. These implants enable a continuous transfer of function from the implant to the regenerating organism. Hence, impairments by remaining implants or surgery for removal can be avoided. [6-9] However, the currently available degradable materials do not meet the requirements of all potential applications. [4][5][10][11] Iron (Fe) - based materials are promising to open up new applications since Iron provides good mechanical performance and is biocompatible. [6][12-15] Nevertheless, the degradation rate of Iron is too low. Alloying with Manganese (Mn) increases the degradation rate, but not sufficiently. [15-19] Further acceleration can be obtained by generating local differences in electrochemical potential inducing anodic dissolution of less noble areas. [20-24] Since FeMn is well-suited, except for its low degradation rate, the material should remain as it is, but the degradation needs to be improved. Thus, the introduction of phases with high potential is promising to modify the degradation of FeMn. For such phases, Silver (Ag) is well-suited as it is biocompatible and has antibacterial properties. In addition, Iron and Silver are almost insoluble in the liquid and solid states. [13][20][21][25-27] The insolubility enables the existence of Silver phases in an unchanged FeMn matrix. However, to fulfill the aim of bioresorbability, the Silver phases have to dissolve after the FeMn. Thus, instead of adding pure Silver, a degradable Silver alloy needs to be developed and applied. Such an alloy requires biocompatibility, which implies an adapted degradation rate to enable an antibacterial effect without harming the organism. Furthermore, the electrochemical potential must be high enough to cause an anodic dissolution of the FeMn matrix.

The insolubility of both components makes the processing of the material combination challenging. [28-29] Powder-based processes enable the mechanical mixing of components as powders and the subsequent processing of the mixture. The powder-bed-based additive manufacturing process laser beam melting (LBM) is well-suited since it offers small melt pools, rapid melting, and solidification as well as strong melt flow. [21][20][30-32] Consequently, immiscible phases of Silver can be created in a FeMn matrix via LBM. Furthermore, the short time at high temperatures reduces the diffusion of alloying elements between the immiscible metallic components.

Therefore, the thesis aims to develop a degradable FeMn-AgX alloy for biomedical applications. Furthermore, it aims to understand the processing of immiscible metals via LBM and the degradation behavior of such LBM-processed materials. The knowledge gained enables the targeted development of materials with the intended properties for the application as Iron-based bioresorbable implants.

2 State of science and technology

The development of materials for bioresorbable implants requires the consideration of manifold requirements and restrictions. Implants need to fulfill biomedical aspects including biocompatibility, adapted mechanical properties, and predictable degradation behavior. These requirements affect each other and need to be fulfilled simultaneously. The addition of some alloying elements e. g. might contribute to the mechanical properties but in turn, might deteriorate the biocompatibility. Since these elements are released during degradation into the human body, compatibility cannot be obtained by an inert behavior and the components have to be selected carefully. Thus, the development starts with the definition of the relevant material properties such as the chemical composition and microstructure to be achieved. Afterwards, the manufacturing process is adapted to achieve the intended properties. Subsequently, experimental investigations are essential to verify whether the intended biocompatibility, degradation behavior, mechanical properties, etc. are obtained. Based on the results, further adaptions are possible. Achieving the intended microstructure is challenging due to restrictions such as the immiscibility of the selected materials. Thus, the restrictions of the processing have to be considered and implemented into the manufacturing strategy for adjusting the intended material. An overall understanding of the manufacturing process is mandatory to enable a targeted adjustment. In the following, the state of the art of the before-mentioned aspects is reviewed as it is the basis for the development of Iron-based degradable materials processed via LBM. Finally, the potential of Iron-based degradable material to reduce patient burden, as well as the remaining challenges, which impede the successful application are addressed with the presented research.

2.1 Degradable implants based on Iron

Advances in medical technology enable the treatment of increasingly more diseases and injuries with implants. A steadily growing number of implants for a wide range of applications are being used to improve the quality of patients` life. [3][4][33] Adapted implants composed of adapted materials are needed to meet the requirements for each application. [1][2][4] The central requirement for materials for implantation is biocompatibility, which is given when a material does not cause adverse consequences such as inflammation, local or systematic toxicity, cancer, allergic reaction, etc.. Currently, the biocompatibility of most implant materials is ensured by inert material properties. Since the development of these materials has been focused on inert properties further potentially beneficial material characteristics such as antibacterial properties or adapted degradation are neglected. [1][4][33-35] Current research is focusing on the unexploited potentials of these materials to enable further applications and improve patients' benefits. An adaption of stiffness can reduce stress-shielding effects (bone growth is not stimulated due to lack of load), antibacterial properties can prevent implant-related infections, and bioresorbable materials can eliminate the need for implant removal surgeries or prevent adverse consequences of implants remaining in the human body. [2][3][6-8][36-38]

In the material development presented, the focus is on bioresorbability, but antibacterial properties are also addressed. Degradability describes the decomposition of the implant. The degradation products can remain. A bioresorbable material degrades in the human body and the remaining degradation products are removed and excreted. [39] Bioresorbable implants are advantageous for applications where implants need to substitute body function for a certain period until the human body can fulfill the task itself again, e. g., an osteosynthesis plate for the fixture of bone fracture or a cardiovascular stent for the treatment of arterial sclerosis. [4][5][9] The duration for which temporal implants are needed depends on the application. Stents should remain for three to six months but not longer than twelve months. After this period the supporting effect is no longer necessary while risks of adverse consequences reveal. [40-42] Since the removal of implants is not feasible in all cases, as, for stents, they can lead to adverse consequences such as in-stent-restenosis (re-closure of a blood vessel within a stent due to tissue growth) or release of alloying elements. Removal of osteosynthesis material, as is commonly conducted, also places a burden on the patient and the overall surgical risk has to be considered. [4][5][9] Thus, a self-dissolving and by the body removable material is beneficial and therefore, in research. Different materials like Iron, Zinc (Zn), and Molybdenum (Mo) are considered degradable materials. A few materials based on Magnesium (Mg) or polymers are already successfully used clinically. [1][6][7][10][39][43] However, these materials do not meet the requirements for all possible applications. The mechanical load capacity of polymer- and Magnesium-based materials is low compared to commercially available inert Iron-based materials. [5][4][10][11] Furthermore, Magnesium-based materials tend to degrade too fast, causing premature loss of mechanical integrity and overload with degradation products accompanied by a significant amount of evolving hydrogen gas. [41][44] Consequently, there is a need for the development of new degradable materials. Iron-based materials are promising due to their biocompatibility, suited mechanical properties, and low degradation rate compared to Magnesium-based materials. [6][10][12-15] Iron is a necessary trace element with a needed daily uptake of 8 mg/d - 27 mg/d, corresponding to a needed uptake of up to 9.8 g/a. [45][46] Accordingly, the amount and period of release of Iron are crucial for biocompatibility. The release of Iron during the degradation of an osteosynthesis plate weighing 2 g or a stent weighing 0.02 g over a few months is uncritical, as the amount required during degradation is higher. However, the release should be continuous and the 2 g must be released over a year and not within a few days. An oversupply by uptake of Iron from the nutrition during the dissolution of an Iron-based implant does not occur. The organism does not absorb Iron from nutrition when the organism is sufficiently supplied. [47]

2.1.1   Degradable implants based on Iron-Manganese

Aside from the biocompatibility pure Iron is not well-suited as the degradation rate is too low and the mechanical properties are reduced compared to commercially Iron-based alloys used for implants. The ferritic structure of pure Iron is ferromagnetic, which impedes the use of magnetic resonance imaging (MRI). [2][6][48][16] A promising approach to overcome these limitations is an adaption of the material properties via alloying. The three mentioned limitations are simultaneously addressed by alloying with Manganese. [12-14][49] Manganese stabilizes the paramagnetic austenitic structure, thus enabling imaging via MRI when the structure is completely austenitic. [16][50][51] Therefore, 17 wt.-% Manganese is necessary when no further austenite stabilizing elements are applied. [52] Furthermore, Manganese reduces the electrochemical potential resulting in an increased degradation rate since materials with less noble character are typically increasingly attacked by corrosion. Segregations of Manganese lead to local differences in the electrochemical potential resulting in an increased anodic dissolution of the less noble areas. This local anodic dissolution further increases the degradation rate. [15-18][53][54] Alloying with Manganese improves the mechanical properties as well. A Manganese content of 30 wt.-% leads to a sufficient reduction of the stacking fault energy to induce the effect of twinning-induced plasticity (TWIP). Hence, TWIP steels exhibit promising mechanical properties, including both, good strength and high ductility. [11][16][20][52][55][56] Application as a material for stents with increased requirements regarding formability and strength is thus also possible. [57][58]

Like Iron, Manganese is an essential trace element, but biocompatibility is more critically discussed. A dose of 4 mg/d, equivalent to 1.5 g/a, should not be exceeded, as Manganese can cause neurodegenerative damage with syndromes comparable to Parkinson's disease. [46][59][60] Therefore, Manganese is not suitable as the base material for degradable implants. [39][61] However, an oversupply of Manganese is mainly excreted and regulated by the liver. [39][61] Thus, an osteosynthesis plate weighing 2 g with a typical amount of 30 wt.-% Manganese containing 0.6 g Manganese is expected to be uncritical. A possibility to reduce the necessary amount of Manganese to achieve a completely austenitic structure and TWIP is alloying with carbon (C). The addition of 1 wt. % Carbon has the same austenite stabilizing effect as 20 wt. % Manganese and up to 1.2 wt. % Carbon can be added. Following this strategy, Iron alloyed with 22 wt.-% Manganese and 0.6 wt. % Carbon showing TWIP is investigated. [20][51][52][56]

2.1.2   Biocompatibility and degradation of Iron-based materials

As indicated above, the amounts of elements released during the degradation should be biocompatible. Nonetheless, the aspect of biocompatibility is more complex than the consideration of single elements and is closely linked to degradation. Ions released during degradation are bound in chemical compounds, the corrosion products. In the complex environment of the living organism, which contains a large number of different elements and chemical compounds, manifold corrosion products are generated. The degradation of Iron-based alloys depends on the composition of the electrolyte, the amount of dissolved gases such as Oxygen (O) which is necessary for the cathodic reaction, and the pH value which is decisive for the solubility of degradation products. [13][23][49][62][63] During degradation corrosion products and further compounds of the surrounding solution can deposit on the surface. These depositions result in a multilayer structure. [62][64][65] These layers influence further degradation of the base material. For example, Phosphates (P) of Iron and Manganese can be present on the surface of Iron-based materials, but they do not inhibit the exchange with the electrolyte. The degradation can proceed further. [15][53][58][62] Ca-P compounds can also deposit on top of the corrosion products. These depositions suppress the contact with the electrolyte and the diffusion of Oxygen. Thus, Ca-P creates an inhibiting layer suppressing further degradation. [14][19][53][56][58][64] The formation of such inhibiting layers has to be prevented to enable a continuously proceeding degradation.

The various degradation products have to be taken into account regarding biocompatibility as well. Degradable FeMn alloys are also investigated in vivo and the results consider good biocompatibility. [14][19][24][53][58][60][66] However, the degradation rate influences the release of material and the concentration of degradation products, which is decisive for biocompatibility. Thus for the adaption of the degradation rate the release of material has to be considered. Biocompatibility is decisive for the successful adaption of the degradation behavior. [15][19][24][52][53][64] The fastest degradation has to enable a sufficiently long supporting effect for healing and ensure a low concentration of degradation products that do not impair biocompatibility. The slowest degradation has to enable dissolution in an acceptable period.

The different conditions in the human organism and the individual response of the human organism have to be considered for adjusting the degradation rate. Challenging might be the individual response of the human organism, as it is not predictable. In case of inflammation, the pH value can reach values down to 5.5. [13][67] Since the degradation depends significantly on the pH value an unpredictable degradation can occur. [68][69] Thus, an adaption of the implant to enable a safe degradation at the intended implantation site for all possible conditions is mandatory. Therefore, degradation at varying pH values has to be considered.

Furthermore, excretion of the degradation products must be possible to meet the requirements of bioresorbability. Hardly dissolvable degradation products remain at the site of implanted Iron-based stents, causing a brownish discoloration of the surrounding tissue. [63][66][70][71] Nevertheless, a complete bioresorption of degradation products from the surrounding tissue can be assumed. Macrophages enable the degradation of the remaining corrosion products as they allow pH values down to 4. Removal by the lymphatic system is expected as degradation products transported to the lymph nodes were observed 52 months after implantation of Iron stents. [13][15][70]

Otherwise, the transport of particles by the lymphatic system within the whole body might be a risk. Wear particles with a size up to 6 μm from implants such as hip joints are detected even e.g. in the liver and spleen. [72-74] Impairments by the displacement of particles to sensitive areas of the human body might be a consequence. The release of elements might be uncritical at the original place of implantation, but not at places degraded particles are transported to, such as the brain. [72][75] Additionally, particles may cause thrombosis, strokes, or tissue damage due to sharp edges. [76][77] Thus, a complete decomposition and excretion of the implanted material and the generated corrosion products is mandatory.

2.1.3   Assessment of biocompatibility of innovative implant materials

Implants and the materials they are made of have to be biocompatible, as mentioned before. The decision of whether this requirement is met might be ambiguous since biocompatibility is not a scalar value with a defined threshold value. The question: "What kind of reactions respectively adverse consequences are acceptable?" has to be discussed critically under consideration of the benefits. If serious issues are solved, minor adverse consequences might be acceptable. [78][79] For example, a slight inflammatory response might be acceptable if an infection is prevented or the implant is degradable and a second surgery can be avoided. [13][14] The prevention of infections is important as they cause high costs and serious consequences for patients, such as repeated operations and increased mortality. Furthermore, ever more heavily treatable infections with multiresistant germs occur. [33][80][81] Accordingly, implants with antibacterial properties might be a great benefit for patients despite some side effects. Due to arguments presented before, the rating of biocompatibility changes. This might enable the application of innovative materials and implants with enormous advantages and only minor side effects. [14][78][79] The changed rating provides new perspectives in research and increases the significance of developing innovative materials for implants with benefits for patients.

The question of biocompatibility is complex, as the response to a foreign body can differ depending on the individual patient. For example, only some humans show allergic reactions to Nickel. [82] Therefore, the decision of whether to use a degradable or conventional implant requires an individualized decision by the patient or the surgeon, based on the individual response of the patient's body and personal assessment of risks and benefits. The research presented aims to develop a material that is suitable for general application. But, even if the material is not suitable for general use, it could be suitable for some applications and provide benefits to patients if they choose such innovative implants. [14][78][79] If a general benefit is not achieved, the efforts of this research could be the basis for application in niche areas.

2.1.4   Modification of Iron-Manganese with Silver for adapted degradation

To summarize the arguments so far: FeMn alloys are promising for application as bioresorbable material. Complete bioresorption and biocompatibility can be expected, especially since changes in the assessment may lead to more materials being declared biocompatible, which is beneficial to patients. Nonetheless, the degradation rate of FeMn is too low for most applications. [15][19][52][53][64] An approach to overcome this limitation is the targeted formation of phases with differing electrochemical potential to force degradation by the anodic dissolution of less noble phases. Since FeMn is appropriate apart from the degradation behavior, the modification of FeMn with phases having a higher electrochemical potential is promising. [20][23][64][83]

Noble elements are well-suited to form such phases in a FeMn matrix as they have a high electrochemical potential. Here, Silver is promising due to its insolubility with Iron, well-investigated biocompatibility, antibacterial effect, and relatively low price compared to Gold (Au), Platinum (Pt), or Palladium (Pd). [21][22][24][84] The complete immiscibility in the liquid and solid state enables the existence of Silver phases in an unmodified FeMn matrix and the properties of well-suited FeMn remain. But the Silver phases affect the properties of the implant structure. The mechanical behavior is influenced since no load-carrying connection between Silver and Iron occurs as they are immiscible. Thus, the Silver phases deteriorate the mechanical properties as if they were pores. The elongation at break and tensile strength are significantly reduced whereas the yield strength is only slightly affected. [20] And, as intended, the modification with Silver enhances the degradation of the matrix material. A successful application of this strategy is reported by various studies. [13][20][21][24-26] Due to the results of Huang et al. [21], Niendorf et al. [20], and Liu et al. [85], an amount of 5 wt.-% Silver is promising as it is a good compromise between enhanced degradation, mechanical properties, biocompatibility, and processability.

As described in section 2.1.2 the formation of corrosion-inhibiting layers is a concern counteracting continuous degradation. The addition of noble phases can contribute to the suppression of the blocking effect of such layers. Wang et al. [62] assume a change in the structure or detachment of inhibiting layers. This hypothesis is confirmed by the increased degradation rates of Silver-modified materials reported in-vivo applications. [13][83] But even the opposite effect is reported as well. [66][86][87] Nevertheless, an increased degradation by modification with Silver is to be expected due to the arguments presented before. However, to obtain a reliable and suitable degradation, the material has to be adapted.

2.1.5 Degradable Silver alloys for the adaption of Iron-based materials

Pure Silver added to FeMn remains after degradation of the matrix. The remaining of Silver is not acceptable, as the aim of bioresorbable implants is complete degradation and excretion to avoid any adverse consequences due to the remaining implant material (cf. section 2.1.2). [88] Therefore, Silver must be modified by alloying to obtain a complete bioresorbable alloy. Such an alloy has to be biocompatible and provide an electrochemical potential high enough to promote anodic dissolution of the FeMn matrix despite the degradability of the Silver alloy. [89][90] Achieving a high potential and degradability are a challenge, as corrosion resistance increases with increasing potential. Consequently, the alloy must enable a trade-off between both aims. In addition, the impact of the released Silver and alloying elements on biocompatibility has to be considered.

Silver enables antibacterial properties, currently utilized e.g. for wound care, but is even discussed controversially as it can cause adverse consequences like argyria. [84][91][92] Argyria describes the irreversible deposition of Silver in organs such as the skin, causing greyish discoloration. For the development of Argyria, the overall intake of Silver is important. [93][94] The release of 0.1 g Silver by an implant weighing 2 g is tolerable, as a lifetime uptake of 10 g is considered non-critical by the World Health Organization (WHO). [95] Aside from the total amount released, the concentration of released Silver ions is decisive, as the mechanisms responsible for the antibacterial effect may act against human cells if the concentration of Silver ions is too high. However, a suitable amount of Silver ions allows antibacterial activity without adverse consequences for the human organism and the adapted release of Silver is promising to produce implants with antibacterial properties. [96-98]

Implant-related infections are an increasing challenge. An exclusion of infections by hygiene precautions alone is not possible, since germs can originate from the patient or may reach the implant by blood circulation after surgery. [3][33][98] Germs that have reached the surface of an implant can form a protective biofilm that makes it difficult for the human immune system and antibiotic therapies to combat them. [3][33] Therefore, a long-term release of antibacterial substances directly from the implant surface, such as Silver, is promising to prevent these infections. Since the antibacterial substances are released where they are needed, the amount required is small. The use of large amounts of antibiotics could thus become obsolete. [3][98] Implants with antibacterial properties are becoming increasingly important, as more infections with multi-resistant germs occur, which harm the progress in the field of implantation. These infections impair progress in the field of implantation and pose a serious threat to patients' life. [4]

However, to enable antibacterial properties an adapted release of Silver has to be achieved by alloying the Silver. The release of Silver ions is in accordance with the intended degradability. Nevertheless, the impact of alloying elements on biocompatibility has to be considered, although the alloying elements are only a small proportion of the degradable implant. The elements selected in this study are uncritical, as Calcium (Ca) is an obvious element with a daily need of up to 1.2 g, and Lanthanum (La) respectively Cerium (Ce) are successfully investigated as alloying elements for degradable Magnesium-based materials. [44][99] The critical values for the uptake of Lanthanum and Cerium are far above the amount released. [100][101] It is therefore assumed that the degradable Silver alloys are biocompatible.

2.2   Processing of immiscible Iron-Silver materials via laser beam melting

The requirements for a degradable Iron-based material result from the previous section 2.1: Homogeneously dispersed phases of a degradable Silver alloy must be embedded in a FeMn-matrix. The generation of such a structure by conventionally manufacturing processes is challenging, as Iron and Silver are immiscible in each other. [28-29] One way to overcome this limitation is the use of powder metallurgical processes, as the compounds can be mixed mechanically. [21][87][102] Aside from sintering, laser beam melting (LBM) is a promising technique. LBM is a powder-bed-based manufacturing process. The powder is melted by laser radiation and the melt pool moves through the material track by track and layer by layer. Thus, the material is consolidated by a continuously proceeding welding process. The process is characterized by a small melt pool, strong melt movements, and rapid melting and solidification, which enables the inclusion of a second, immiscible phase. [20][30-32] Other advantages of additive manufacturing include the increased design freedom and the possibility to manufacture individually adapted implants. [103][104] Accordingly, LBM is a promising technic to fabricate Iron-based degradable implants modified with Silver. Nonetheless, for a successful application, the morphology and chemical composition of the Silver phases must be controllable as they are crucial for biocompatibility, degradation behavior, and mechanical properties. For instance, nanoparticular Silver is discussed critically in the literature and should be avoided. [32][105] Apart from that, the phases should be small and well dispersed to enable homogenous degradation and minimize the influence on the mechanical properties. However, lower stiffness may be useful to match the stiffness of the human bone and reduce stress shielding. Adjusting stiffness by creating a targeted porosity or scaffolds via LBM is investigated. [36][106] In addition, a targeted porosity can be applied to adapt the degradation behavior and improve osteoconductivity (bone growth on a surface). [14][107] Following this strategy, the formation of a targeted porosity could be useful in addition to the modification with Silver.

In summary, the properties of the implant are influenced by the structure resulting from the LBM process. Hence, a fundamental understanding of the LBM process is mandatory to allow a defined adaption of the raw material and the LBM process to obtain a material with the desired properties. This requires an in-depth analysis of the melt pool and the interaction of the two immiscible phases.

2.2.1 Approach for analysis and modeling of the laser beam melting process

Obtaining knowledge of the influence of physical effects on the continuous formation of bulk material during the LBM process is the subject of current research. This knowledge enables targeted adaption and avoids extensive random testing of different process strategies. [108-111] In addition to the LBM process strategy, the feedstock properties like chemical composition, the proportion of immiscible materials, and the morphology of the powders influence the final material properties. These input parameters can be used to adjust the material properties, e.g., vaporization of alloying elements can be compensated by oversizing the raw material. [112][113]

Although the fundamentals of the LBM process are always identical, the specific characteristic of LBM processes differ significantly depending on the processed material, powder properties, and applied LBM parameters. [111][114] Understanding the physical effects is further complicated since the effects cannot be observed directly. Knowledge can be gained by observing the melt pool, transferring known physical effects, and investigating the microstructure and properties of the final material. Based on these findings, models are generated to simulate the melt pool and LBM process. Initially, in the development of such models, only the basic influences are considered. Subsequently, detailed simulations are developed in which several physical effects are taken into account. The models and underlying assumptions of the LBM process are verified by experimental validation. For the simulation of the LBM process, satisfying results are obtained. [108-110][115][116]

However, many models only deal with single tracks, and varying conditions caused by previously built adjacent tracks and irregular structures of previous slices are not considered. Accordingly, the real melting process is less stable than the simulated one. [117-119] Nevertheless, elementary effects can be investigated, and it is possible to gain an understanding of the influence of physical effects. The models allow a specific adaption of the LBM process.

However, models considering two immiscible metallic materials processed via LBM are not available so far. Hence, such a model needs to be developed. The basis for developing such a model is the knowledge available in the literature. But this knowledge must be adapted to the LBM process with two immiscible phases. This requires correlation with the observed microstructure and mechanical properties of the Iron-based material modified with Silver. The mechanisms decisive for the formation of Silver phases as well as the interaction between both immiscible components with each other are determined. Hence, the model is developed following the approach presented in the literature. In the following sections, the state of the art of the melt pool and the physical effects influencing the melt pool are presented.

2.2.2 Physical effects decisive for the formation and characteristic of the melt pool

The successive formation of bulk material during LBM is the result of an interaction of sequential proceeding sub-processes respectively physical effects caused by laser radiation. The main processes are absorption of laser energy, melting, consolidation of the molten powder particles, melt flow within the melt pool, heat dissipation, and solidification. These processes occur continuously while the laser is moving along the track. [108][110][111][115] These sub-processes interact with each other and cause further effects like spattering and chemical reactions. The characteristics of the sub-processes depend on further physical effects e. g. the spattering is influenced by the vaporization of the material. [111][115][120-122] Accordingly, multiple material properties, process conditions, and interacting physical effects are responsible for the characteristics of the melt pool and the resulting structure of the LBM-processed material.

An example of the interaction of physical effects is the absorption of laser radiation: The formation of the melt pool starts with the absorption, which depends on the absorptivity of the material in the molten and solidified state. In addition, reflection in the powder bed, a deepening of the melt that may occur, and deflection by the vapor plume can affect the amount of energy absorbed. Which proportion of the laser radiation reaches the powder bed or the melt influences the absorption as well. [123-126]

However, there is consistency in the literature on the predominantly influencing physical effects, although the reported melt flow with significant influence on the resulting shape of the melt pool differs. [108-110][115][124] The main influencing effects are the surface tension resulting in Marangoni flow and evaporation of material under the laser spot which results in recoil pressure. [108][116][123][127][128] Further less important influencing effects are the gravity, buoyancy force, drag force, thermal conductivity, heat radiation, dynamical wetting, and eventually occurring phase transformations. [108][110][116][127]

The intensity of evaporation is the decisive factor determining the shape of the melt pool. The shape of the melt pool is an indication of the melt flow and the welding mode. Three welding modes are distinguished: When there is only low recoil pressure, the surface of the melt pool is almost flat and heat is transferred into the material mainly by conduction. Accordingly, the mode is defined as conduction mode. [108][124][129][130] When the recoil pressure is high, the molten material is displaced and a deep and narrow depression, a so-called keyhole, develops. This welding mode is referred to as keyhole welding mode. The laser radiation penetrates the material and heat is transferred primarily by the convection of vaporized material. [108][124][129][130] The formation of a keyhole starts with the development of a depression. Therefore, the recoil pressure of the vaporized material must be high enough to displace molten material against the surface tension. Additional surface energy is required to form the additional surface of the depression. [131-133] As the laser is reflected within a depression it further increases energy absorption and more material evaporates. The depression supports itself and can develop into a keyhole when a threshold value of energy input is exceeded. [131][134][135] Inside a keyhole the laser is reflected multiple times increasing the energy absorption and vaporization. As a result, a keyhole stabilizes itself and the size of the melt pool increases. [131][136-138] The addition of alloying elements with high vapor pressure and distinct vaporization supports the formation of a keyhole. Thus, such elements can lower the threshold of required energy input for keyhole welding mode. [132][133][135][138] Accordingly, the formation of a keyhole depends on both the material properties and the processing parameters.

Since the transition between conduction and keyhole mode is continuous, with a differing proportion of characteristics of both modes, a so-called transition mode is defined. [108][124][129][130] Depending on the material and process parameters all modes occur in LBM. For Iron-based materials, the transition mode is most common to produce material with suitable properties. [121][125][129][130]

Evaporation is therefore crucial for the formation of the melt pool in the interaction zone with the laser. However, the melt pool also extends beyond the interaction zone with the laser, and other effects are important in this zone. Following Khairallah et al. [108], the melt pool can be divided into three sections. The first part of the melt pool before and in the region of the laser spot is mainly determined by the recoil pressure in the keyhole and transition mode. Thus, evaporation and welding mode are crucial for the formation of the Silver phases, as the recoil pressure determines the melt flow in the interaction zone with the laser. Although different vaporization behavior of the two immiscible phases is expected, the resulting vapor above the melt pool should lead to an identical pressure effect on both phases.

The transition and tail-end region are following the interaction zone with the laser and are predominantly prevailed by surface tension. [108] The surface tension depends on the temperature, concentration of alloying elements and amount of absorbed oxygen, etc. Differences in surface tension cause a flow of melt from low to high surface tension, the so-called Marangoni effect. [139-142] Thus, the surface tension defines the melt flow and surface structure of the solidified melt tracks. For the conduction mode, the surface tension is decisive in the first part of the melt pool, too. [108] As the shape of Silver phases is defined during solidification and the position results from former melt flow, the transition and tail-end region are decisive for the morphology of the Silver phases. Thus, the formation of the Silver phases is mainly influenced by the surface tension and the resulting Marangoni flow. Accordingly, the surface tension and in particular the differences in surface tension are crucial for analyzing the formation of Silver phases and developing a corresponding model.

However, besides the physical properties, the reactivity respectively the chemical properties of the elements influence the LBM process, too. In particular, elements with high reactivity can react with each other or with oxygen since it is not possible to completely remove the oxygen from the building chamber. [117][143][144] The reactivity increases with increasing temperature. In this way, distinctive chemical reactions can form oxide layers deteriorating the connection between adjacent and superimposed melt tracks. [117][143][145][146] Furthermore, the mechanical properties can be affected by slag incorporated into the bulk material. The slag is typically located on the top of the melt tracks. [143][147][148]

2.2.3 Material and process-dependent spattering affecting the composition

Since spattering can influence the microstructure and composition of a mixed material processed via LBM, this effect is described here in detail. Various effects are responsible for the formation of different kinds of spatter. [115][121][149][150] Ejections of liquid material from the melt pool mainly result from recoil pressure and Marangoni forces respectively instabilities of the melt flow. [121][126][151] Melt ejected with high velocity solidifies in flight and forms spherical particles. Whereas melt ejected with a low angle and low velocity reaches the powder bed before solidification. When the melt reaches the powder bed, it can enclose multiple powder particles and form agglomerations. [115][149] Most spatters are generated at the melt pool front due to displacement by the forward-moving melt pool front and evaporation. [108][121][124] When present, a keyhole significantly contributes to spattering as the material is displaced and accelerated by vapor. [115][121][150] As the recoil pressure decreases beyond the interaction zone with the laser, the surface tension causes the keyhole to close. The melt required for closure is accelerated from the front of the keyhole around it to the back. [121][123] Some of this melt flowing around the melt pool can be sheared off on the upper rear forming welding spatter. The intensity of spattering caused by this effect depends on material properties such as the viscosity of the melt and the surface tension. [115][121][151] Accordingly, the tendency to generate ejections of immiscible components might vary. Preferential ejection of one single component might occur which changes the composition of the material. Therefore, it might be necessary to compensate for the preferential spattering by overweighting this component.

Aside from the material properties, process conditions such as energy density, scanning velocity, and gas flow also affect spattering. [115][150] Hence, spattering has to be taken into account for the adaption of the LBM process, too.

Beyond the ejection of melt, even non-molten powder is ejected. The heating and vaporization at the center of the melt pool create an upward gas flow causing an inward gas flow at the sides of the melt pool, which entrains powder particles from the adjacent powder bed. This entrainment of surrounding powder can lead to a powder-free denudation zone along the melt track. The powder entrained with the gas flow is only partially ejected as it is also incorporated into the melt pool. [120][121][124][152] However, the attraction of surrounding powder by the melt pool contributes to the denudation zone. [108][120]

Another effect can result in the ejection of powder particles from the powder bed. Rapid heating of the surface of powder particles can cause evaporation of the surface material, creating a bow-off impulse that accelerates powder particles. [115][124][150][153] Since evaporation depends on the material properties, this effect might contribute to the selective ejection of one of the two immiscible components.

2.2.4   Diffusion of alloying elements affecting immiscible phases

One further effect must be considered when processing immiscible materials via melting processes: The diffusion of alloying elements from one metal into another liquid metal. This effect can be very pronounced and is even used e. g. to separate metals for recycling or to obtain porous structures. [154][155] However, for the formation of a FeMn-based matrix with degradable Silver phases, the diffusion of alloying elements is not desired. Both, the matrix and the degradable Silver alloy, should not be affected by changes in the chemical composition. In this context, processing via LBM might be advantageous as melting and solidification occur very fast, leaving little time for the diffusion of alloying elements between both melts.

However, changes in the chemical composition of both components might occur during LBM. A diffusion of alloying elements of the degradable Silver alloy into the matrix would create a deficiency of these elements in the Silver phases deteriorating the degradation behavior. In contrast, the uptake of these elements by the Iron-based matrix is not critical for the selected elements. Calcium and Lanthanum are used as deoxidizing agents. [156-159] Both elements are soluble in Iron in the molten state but nearly immiscible at room temperature. [27] These elements tend to react with oxygen or alloying elements and form inclusions, which can cause grain refinement or improve toughness. [159][160] Accordingly, diffusion of Calcium and Lanthanum from the Silver into the Iron is acceptable as long as the Silver phases remain degradable.

A diffusion of Manganese from the Iron-based matrix into the Silver reduces the Manganese content within the matrix material, which might affect TWIP or the austenitic structure. However, only 6 wt.-% Manganese is soluble in Silver at room temperature [27] and the combined Manganese and Carbon content is high enough, that even a loss of 2 wt.-% Manganese would not change TWIP or the austenitic structure. [52] A negative influence on the properties of the matrix is therefore not to be expected. Furthermore, the uptake of Manganese by Silver could contribute to the degradability of the Silver phases. Or else, the diffusion of Manganese might reduce the electrochemical potential of the Silver phases and affect the intended anodic dissolution of the Iron-based matrix.

To sum up, the diffusion of alloying elements is not critical for the properties of the matrix material but might cause concerns regarding the Silver phases, since the degradation behavior and electrochemical potential might be affected. Accordingly, the adjustment of the LBM process aims to reduce the diffusion between both components. The intensity of interaction depends on the temperature and the duration of high temperature during LBM. For this reason, a reduction of the energy input should be favorable to reduce the diffusion and preserve the alloying elements. Nevertheless, the aim of a suitable morphology of the Silver phases must be addressed simultaneously.

3 Motivation and research hypothesis

The fundamental motivation for this thesis is to improve the quality of life of patients suffering from diseases and injuries that could be treated with bioresorbable materials but are treated with less-suitable inert implants due to the lack of suitable bioresorbable materials. Iron-based materials are promising to fulfill the requirements for some of these applications, but defined modifications are required. As mentioned before, FeMn modified with a degradable Silver alloy and processed via LBM is highly promising as bioresorbable material with antibacterial properties. However, this strategy is also critically discussed due to remaining concerns. [19][66][86][161][162] Thus, further investigations and adaptions are required to enable safe and appropriate degradation and to prove biocompatibility. Therefore, a fundamental understanding of how LBM affects the microstructure and how the microstructure influences the resulting material properties such as degradation behavior and mechanical properties is mandatory. Knowledge of these correlations will allow the required microstructure to be defined and, in the next step, to adapt the manufacturing process and raw material to achieve the desired microstructure. The research presented is intended to enable this strategic development.

Accordingly, the following research hypothesis is addressed:

> FeMn modified with inclusions of a degradable Silver alloy can be used as bioresorbable material with suitable mechanical properties, adapted degradation rate, biocompatibility, and antibacterial effect.

This hypothesis is addressed by the six publications included in this thesis. The hypothesis is divided into sub-hypotheses:

a) Alloying of Silver allows the creation of a degradable and biocompatible Silver alloy with an open circuit potential that is high enough to evoke the anodic dissolution of FeMn. (Addressed in publications A, B.)

b) LBM enables the formation of the desired phases of a degradable Silver alloy in a FeMn matrix. The formation of these Silver phases is based on regular and controllable processes so that a model of the formation enables a specific, targeted adaption. (Addressed in publications C, D, E.)

c) Phases of a degradable Silver alloy evoke anodic dissolution and accelerated degradation of the FeMn matrix. (Addressed in publications E, F.)

d) FeMn modified with AgCaLa is biocompatible and has antibacterial properties. (Addressed in publications A, F)

Contents lists available at ScienceDirect

Journal of Alloys and Compounds

journal homepage: www.elsevier.com/locate/jalcom

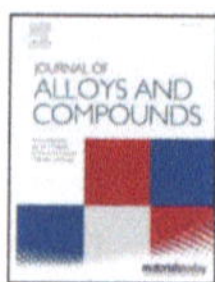

Novel AgCa and AgCaLa alloys for Fe-based bioresorbable implants with adapted degradation

Jan Tobias Krüger[a,*], Kay-Peter Hoyer[a], Viviane Filor[b], Sudipta Pramanik[a], Manfred Kietzmann[b], Jessica Meißner[b], Mirko Schaper[a]

[a] Chair of Materials Science, Paderborn University, Mersinweg 7, 33100 Paderborn, Germany
[b] Department of Pharmacology, Toxicology and Pharmacy, University of Veterinary Medicine Hanover, Foundation, Buenteweg 17, 30559 Hanover, Germany

ARTICLE INFO

Article history:
Received 3 November 2020
Received in revised form 18 February 2021
Accepted 13 March 2021
Available online 19 March 2021

Keywords:
Silver alloys
Corrosion
Biomedical application
Laser beam melting
Bioresorbable metal
Cytocompatibility
Iron-based alloys

ABSTRACT

Resorbable implants are highly beneficial to reduce patient burden since they need not be removed after a defined period. Currently, magnesium (Mg) and polymers are being applied as bioresorbable materials. However, for some applications the insufficient mechanical properties and high degradation rate of Mg cause the need for new materials. Iron (Fe)-based alloys are promising due to their biocompatibility and good mechanical properties, but their degradation rate is too low and needs to be adapted eg. via alloying with manganese (Mn). Besides, phases with high electrochemical potential lead to increased degradation of residual material with lower potential based on the galvanic coupling. Here, silver (Ag) is promising for the formation of such phases due to its high electrochemical potential (+0.8 V vs. SHE), immiscibility with Fe, biocompatibility, and anti-bacterial properties. Since remaining silver particles can lead to adverse consequences as thrombosis, these particles need to dissolve after the matrix material. Thus a silver alloy with high electrochemical potential, biocompatibility, and adjusted degradation behavior is required as an additive for iron-based bioresorbable materials. Several silver alloying systems are possible, but regarding the electrochemical potential and degradation behavior of binary alloys, calcium (Ca) and lanthanum (La) are best-suited considering their biocompatibility. Accordingly, this research addresses AgCa and AgCaLa alloys as additives for iron-based degradable materials with adapted degradation behavior

© 2021 Elsevier B.V. All rights reserved.

1. Introduction

Iron (Fe)-based alloys are in focus as bioresorbable materials for implants, such as stents or osteosynthesis plates [1–4]. Since iron is essential for the human body and shows only negligible toxicity it fulfills the requirements for biocompatibility [5,6]. Besides, the degradation rate is moderate, so the degradation rate is not needed to be reduced as for Mg-based alloys [1]. Also advantageous regarding the degradation in comparison to Mg-based degradable alloys is the reduced generation of gaseous corrosion products [7,8]. Furthermore, iron is characterized by superior mechanical properties, accordingly especially load-bearing implants can be designed with more filigree and less material is inserted in the human body. This, in combination with the slow degradation of iron-based alloys, is promising for the addressed application [9,10]. However, a modification of these alloys is necessary due to the insufficient

degradation rate, as iron-based materials can show properties that are a prerequisite for permanent implants [1,2,6]. Furthermore, the high stiffness compared to the human bone needs to be adapted regarding disadvantageous bone growth and stress shielding [11]. Hence, iron-based alloys with balanced properties have to be developed.

An increasing degradation rate can be achieved by alloying iron with manganese (Mn). Two effects are possible; on the one hand, manganese leads to a more negative electrochemical potential, on the other hand, phases with different potentials evolve [12–14]. Furthermore, the addition of manganese results in better plasticity combined with good strength, which is advantageous for stents that need to be formable [9,14,15]. Further benefits of iron alloys with adequate manganese content are the paramagnetic properties. Hence magnetic resonance tomography (MRT) can be applied for the post-implantation diagnostic [14,16]. Due to the formation of passivating layers, the degradation rate of high manganese steel still needs to be increased [12,13,17]. The formation of local areas with different electrochemical potentials enhances the material dissolution, based on the formation of galvanic cells and the resulting

* Corresponding author.
  E-mail address: krueger@lwk.upb.de (J.T. Krüger).

https://doi.org/10.1016/j.jallcom.2021.159544

anodic dissolution of the material with a less noble potential [18,19]. Areas with different potentials can be achieved by different phases, respectively by the formation of a multi-phase structure. Subsequently, to further accelerate the degradation of iron-manganese alloys, phases with a noble potential need to be generated. Therefore, iron insoluble elements with noble potential can be selected or the formation of appropriate intermetallic phases can be promoted. Consequently, the addition of elements with higher electrochemical potential should increase degradation, and elements with the highest possible potentials, i.e., precious metals, are best suited. Thus, different precious metals are examined as an additive for iron-based biodegradable materials. Here, Huang et al. [19] showed that silver, gold (Au), palladium (Pd), and platinum (Pt) act as effective cathodes to increase the corrosion rate of Fe-based alloys. Similar effects are reported for the addition of palladium to high manganese steel [12,17]. Accordingly, a promising approach is to add silver, as it is biocompatible and anti-bacterial [21]. Silver should increase the degradation rate due to its positive electrochemical potential as the difference in potential referred to iron is 1.32 V [18,22]. The effectiveness of silver is confirmed by various studies showing enhanced degradation rates and a more uniform degradation for iron alloys containing silver phases [18,20,22,23].

The immiscibility of elements like iron and silver on the one hand promotes the evolution of phases with high electrochemical potential, but on the other hand, makes the processing of such an alloy challenging. In this regard, new manufacturing methods have to be considered, which allow the processing of conventionally immiscible material systems. With additive manufacturing it is possible to overcome these disadvantages and also, to allow unique processing conditions, like unprecedented design-freedom and individualized geometries [24,25]. In previous work, the processability of a Fe-Mn-Ag alloy via laser beam melting (LBM) was proven [18,22]. The mechanical properties, as well as the corrosion behavior of the prepared alloys, are promising. The silver phases remain after the dissolution of the surrounding iron-based matrix. In terms of bio-medical applications, the release of silver must be addressed, as an enrichment of silver in the biological environment should be avoided. The influence of silver is discussed controversially as both, the advantages and disadvantages have to be taken into account [26]. Silver acts antibacterial and is already used in medicine for example for the therapy of wounds and in the past for the therapy of diseases like syphilis [26,27]. Apart from that, a large intake of silver can lead to argyria [27,28]. Due to these effects, the amount of silver intake is the key factor for biocompatibility. For example, a stent with a weight of 0.02 g or an osteosynthesis plate with a weight of 2.0 g made of an iron alloy containing 5.0 wt.-% silver would release 0.001 g respectively 0.1 g silver. This amount of silver is tolerable as the World Health Organization (WHO) proposes that a lifetime intake of up to 10 g silver is uncritical [29,30].

Nevertheless, the silver phases released during the degradation of the implant in the tissue may lead to adverse consequences as thrombosis or tissue damage due to sharp edges. Silver particles, especially nanoparticles, are critically discussed in the literature [31,32]. Therefore, the silver phases have to be modified in terms of biocompatibility and self-degradation to enable the phagocytosis of the released particles.

To identify the most promising alloying concepts for degradable silver alloys, various binary silver alloys containing magnesium (Mg), calcium (Ca), manganese (Mn), zinc (Zn), germanium (Ge), lanthanum (La), cerium (Ce), neodymium (Nd) and/or ytterbium (Yb) under conditions similar to the conditions presented in this work have been investigated in preliminary work (unpublished research). In preliminary work [33], first promising alloy systems with calcium have been presented. Alloys with 6.0 wt.-% and 10.0 wt.-% calcium show promising degradation behavior [33]. Based on the low corrosion resistance of elementary calcium and the presence of different intermetallic phases such as Ag_2Ca and $AgCa$ with different electrochemical potential, the Ag-Ca-system was addressed [33,34]. Calcium is an obvious element due to good biocompatibility as it is vital for the human body and a daily need of 0.8 g up to 1.2 g is generally accepted [35].

However, silver-calcium alloys show comparably inhomogeneous degradation due to the coarse microstructure of the different phases, and the high degradation rate decreases over time. In contrast, the relatively high degradation rate of binary lanthanum alloys increases over time, and a uniform degradation on the entire surface was noted in previous work. For the AgLa alloys, high degradation rates of ≈ 90 g/m^2 have been measured during seven weeks of immersion in Ringer's Lactate solution. Hence, by adjusting the lanthanum content the degradation rate is significantly affected. Therefore, in theory, the combination of both binary alloys would lead to a promising ternary silver alloy. Although the superimposition of the impact of single alloying elements is not recommended, alloying of silver with calcium and lanthanum seems to be a promising approach due to the immiscibility of calcium and lanthanum [34]. The immiscibility might promote the formation of phases containing calcium and those containing lanthanum. These phases should have the same properties as in the binary system resulting in adjustable, adapted degradation behavior.

Rare-earth elements, such as lanthanum, are not essential for the human body and their impact on the human organism is discussed controversially [36]. As the implants and accordingly the lanthanum is directly introduced to the tissue examinations of injected lanthanum into the tissue or under the skin are considered here. Regarding the related atomic settings of rare-earth metals, the properties of these elements are similar. For intravenous injection of rare-earth metals, LD50 values between 10 mg/kg and 100 mg/kg are reported. The LD50 value for the intraperitoneal injection ranges from 150 mg/kg to 700 mg/kg [37]. Injection above 100 mg/kg under prolong administration could e.g. lead to liver damage [37]. But otherwise, chemical compounds of these elements can be excreted quickly and are therefore less critical to the human body. Accordingly, e.g. lanthanum carbonate can be used as a phosphate binder for dialysis patients [38]. Additionally, lanthanum is already examined as a component of magnesium-based degradable materials. In vitro and in vivo tests of a magnesium alloy containing 0.69 wt.-% lanthanum lead to promising results regarding biocompatibility [39]. Since the expected amount of lanthanum in the final FeMnAg (containing 5 wt.-% of the AgCaLa alloy) will only be 0.05 mg for a stent with a weight of 0.02 g and 5 mg for an osteosynthesis plate with a weight of 2 g, the introduced amount of lanthanum is expected to be uncritical regarding the mentioned critical values. Furthermore, the release of lanthanum over a long time should promote tolerance.

Overall, it can be stated that the considerations listed above are quite strong and provide an indication of the choice of alloys with the greatest potential from the current state of the art. A FeMnAg alloy processed via laser beam melting can address various issues: Sufficient mechanical properties, an increased degradation rate, and good biocompatibility. To make this concept successful, a degradable silver alloy is needed as an additive for iron-based alloys. Therefore, the present study focuses on AgCa and AgCaLa alloys. These calcium and lanthanum containing silver-based alloys should combine the advantageous properties of the individual elements, determined by preliminary work. As a reference, the degradation rate of a binary AgCa alloy was determined.

Both alloys were melted in a graphite crucible in an inert gas atmosphere, using argon gas. An evaluation of the microstructure was undertaken by scanning electron microscopy (SEM) coupled with energy dispersive spectroscopy (EDS). The phase formation was characterized by cooling curves, SEM, and X-ray diffraction (XRD) scans. Controlled degradation of the alloys was performed by immersion testing under conditions imitating the human body and intermittent weighing of the samples. To determine the biocompatibility cell viability and proliferation were examined. The targeted correlation between the content of alloying elements with the developing microstructure and the resulting properties regarding degradation behavior and biocompatibility should enable further directed adjustment of the alloying system.

2. Experimental procedures

2.1. Alloy production

The AgCa and AgCaLa alloys were produced by conventional melting in a graphite crucible, using a conductive furnace (Goldbrunn 1000) and subsequently cast into a graphite mold. Both, the crucible and the mold were coated with boron nitride to increase the service life and to guarantee a better unmolding. Furthermore, boron nitride does not interact with AgCa alloys [40]. To prevent oxidization and air ingress, the crucible was rinsed with argon gas. This prevents the reactive alloying elements calcium and lanthanum from burning and silver from absorbing oxygen. To ensure protection from oxygen during casting, the crucible opening was safeguarded with an oxygen burning flame. The burner flame prevents the melt from cooling too quickly. Furthermore, the molds were heated to ≈ 700 °C before casting, to reduce thermal shock.

For alloy production, the following procedure was employed: Initially, the silver, calcium, and lanthanum granules were weighed with a precision scale (PCE-BT 200) with an accuracy of 4.0 mg. The used elements have a purity of 99.99% (Ag), 98.5% (Ca), and 99.9% (La). After the melting of silver, the alloying elements were added at a temperature of 1040 °C. This temperature was chosen as it has been proven in prior investigations, that this is a good trade-off between fast melting of the alloying element and a reduction of melting loss. Nevertheless, for the AgCa and AgCaLa alloys, a smoke emission was observed despite shielding by the protective gas. This suggests that either small remaining amounts of oxygen reacted within the crucible or evaporated calcium reacted with the atmospheric oxygen after exiting the furnace. Hence, the calcium content diminishes during casting. To compensate for the burn-off, the amount of calcium was extended based on experiences with previous alloys. Besides, the time between the additions of the alloying elements until casting was kept as short as possible but to ensure homogenous mixing. Subsequently, coarse and loose oxides were removed from the cast block.

To record the temperature, thermocouples (Ni/CrNi, Type-K) were connected to a temperature-measuring module (DAQmx, National Instruments) and the measured values were recorded with the software antoha 2019. The current lines are inserted through holes into the mold and connected via spot welding. The prepared thermocouples were cast in and the data were collected at a frequency of 50 Hz from the filling of the mold until the cast block reached 200 °C. This measuring procedure has been validated with reference measurements in which the well-known melting point of pure silver has been identified with an accuracy of +/- 2 °C.

2.2. Sample preparation

The samples were wire eroded from the as-cast blocks (14 × 14 × 36 mm³) into vertical slices with a dimension of 10 × 10 × 2 mm³. For further preparation, the layer from erosion was first removed by grinding. A water-free suspension based on ethanol was used to prevent corrosion. Subsequently, all samples were cleaned and dried with ethanol. In the last step, all samples were polished applying water-free abrasives of grit sizes 3.0 μm and 1.0 μm.

2.3. Microstructural investigations

2.3.1. X-ray diffraction (XRD)

To detect and specify the evolved phases, XRD was performed (X'pert, Pan Analytical) applying a power of 2.2 kW. A graphite monochromator was employed and XRD was conducted using Cu-Kα radiation. For measuring, the XRD goniometer was operated in line focus mode and the samples were scanned from 20° to 120°. Indexing of the XRD peaks was done using XPert Highscore and MAUD [41].

2.3.2. Scanning electron microscopy (SEM) and energy dispersive spectroscopy (EDS)

SEM (Zeiss Ultra Plus) was conducted to analyze the microstructure of the alloys. To determine the chemical composition as well as to depict the element distribution, the alloy composition was investigated via energy dispersive spectroscopy (EDS). EDS maps were acquired using an EDAX AMETEK detector.

2.4. Corrosion measurements

2.4.1. Open circuit potential measurements (OCP)

Open circuit potential measurements (OCP) were conducted with the Potentiostat MLab200 (Bank electronics) to monitor the free corrosion potential. An Ag/AgCl electrode in a saturated KCl solution (Meinsberger, Typ SE11) was taken as a reference electrode and a platinum sheet as a counter electrode. The potential of the reference electrode at 25 °C deviates by 198 mV from the potential of the standard hydrogen electrode (SHE) and the measured values have to be corrected. Before the measurements, all samples were ground to remove any oxide layers. The samples were fixed with a conductive clamp and dipped in the electrolyte with a surface area of 140 mm². Concerning the addressed application, Ringer's Lactate solution, containing 131.0 mmol/l Na, 5.4 mmol/l K, 1.8 mmol/l Ca, 112.0 mmol/l Cl and 28.0 mmol lactate, was taken as an electrolyte. Ringer's Lactate solution contains the significant electrolytes of human blood in similar concentrations. After arranging the test set-up, 5 min were given to achieve a state of balance. The measurement of the voltage was then started at a recording rate of 10 Hz for 30 min.

2.4.2. Immersion test

Immersion tests were performed to investigate the time-depending degradation behavior of the AgCa and AgCaLa alloys over a long period. In the test, replication of the human body conditions was undertaken. Therefore, Ringer's Lactate solution was used as the electrolyte and filled into small plastic bottles. Three samples of each alloy were tested in one bottle with a volume of 100 ml. Besides this, air was blown with a rate of 12.5 l/h into the solution in each bottle to allow the exchange with the ambient air to imitate the exchange through the lungs and to ensure a stable concentration of dissolved gases, especially oxygen. The electrolyte temperature was raised to 37 °C and kept constant during the whole test duration, i.e., to simulate body temperature. At last, the bottles were placed on a laboratory shaker operating at low frequency to archive homogenous conditions all over the samples and provide differences in concentrations of electrolyte. For hanging up the samples in the solution small holes were drilled into the specimens. For all components of the experimental setup in contact with the solution, inert synthetic materials were used.

All samples were weighed regularly on a precision scale with an accuracy of 0.01 mg (XP205, Mettler Toledo GmbH) to determine mass changes throughout the test period. The time between the weighing was kept short at the beginning due to the fast degradation and then extend to a weekly control. For weighing, each sample was removed from the electrolyte and rinsed with a cleaner for Ringer's

Lactate solution followed by rinsing with acetone and drying. Ringer's Lactate solution was replaced with a new solution. As some solution evaporates within a week due to the airstream, the liquid level was regularly supplemented by the addition of Ringer's Lactate solution. Additionally, the pH-values of the degradation solutions were measured always before the solutions were renewed.

2.5. Biocompatibility investigations

2.5.1. Cytocompatibility tests

The experiments were carried out with the murine fibroblast cell line L-929 obtained from CLS Cell Lines Service (Eppelheim, Germany) and the human osteosarcoma cell line HOS 8707020287,070,202 (HOS) obtained from the European Collection of Authenticated Cell Cultures (Merck, Germany). L-929 cells were grown and passaged in RPMI-1640 medium (Biochrom GmbH, Berlin, Germany) supplemented with 10% fetal calf serum (FCS) superior (Biochrom GmbH) and 1% penicillin/streptomycin (Biochrom GmbH). Eagle's MEM (EMEM)/Hanks' (Carl Roth GmbH + Co. KG, Karlsruhe, Germany) maintained 10% FCS, 1% penicillin/streptomycin, 1% non-essential amino acids (Biochrom GmbH), and 2 mmol L-glutamine (Biochrom GmbH) was used for HOS cells. Both cell lines were grown and passaged in cell culture flasks or multiwell plates purchased by Greiner Bio-One, Frickenhaus Germany. A 0.05% trypsin/0.02% ethylene-diamine-tetraacetic acid solution (Biochrom GmbH) was used for passaging. The cells were plated with a density of 10,000 cells per well in 96-well culture plates. Cells underwent 24 h of incubation with the degradation medium in a humidified atmosphere at 37 °C and 5% CO_2. Therefore, the samples ($4 \times 4 \times 2$ mm^3) were incubated in 1 ml, 3 ml, or 10 ml of RPMI-1640- or EMEM-medium, without additives for 3 days in an incubator at 37 °C and 5% CO_2. The cells were subsequently treated with undiluted degradation medium, a dilution of 1:10, and a 1:100.

2.5.2. Cell viability

Cell viability was determined by MTS assay (CellTiter 96® Aqueous One Solution Cell Proliferation Assay, Promega, Madison, USA) with confluent cells, after incubating with degradation media for 24 h according to the manufacturer's protocol. In brief, the supernatant containing the degradation medium of the samples was removed from the cells. Afterward, 100 µl of CellTiter 96® Aqueous One Solution Reagent in RPMI-1640- or EMEM-medium (1:6) were added to each well of the 96-well array plate and the plate was incubated at 37 °C for 4 h in a humidified 5% CO_2 atmosphere. To measure the amount of soluble formazan produced by the cellular reduction of MTS, the absorbance was determined at 490 nm using a multiwell plate reader (Dynatech Laboratories, Denkendorf, Germany). Each concentration was measured four to eight times, while each experiment was performed 6 times. Background absorbance (490 nm) was corrected by using control wells containing the same volumes of culture medium and CellTiter 96® Aqueous One Solution Reagent.

2.5.3. Cell proliferation

The inhibitory effect of the degradation medium was examined using a crystal violet staining (CV) assay [42]. In brief, cells were seeded into a 96-well microtiter plate with a density of 10,000 cells per well to avoid confluence of cells during the experimental period. After cell adherence (4 h), cells were treated with serial dilutions of the different degradation media. Controls were only treated with the medium. The proportion of treated cells compared to untreated controls was determined 24 h after the exposure. After that time, the cells were fixed with 2% glutaraldehyde (Sigma-Aldrich) in phosphate-buffered saline (PBS) for 20 min. After removal, cells were stained with 0.1% crystal violet (Roth GmbH) in deionized water for 30 min. The plates were air-dried after washing with deionized

water. Crystal violet was then dissolved out of the cells by adding 2% Triton X-100 (Sigma-Aldrich, Steinheim, Germany) in deionized water. Following 1 h incubation, the absorbance at 570 nm was determined using a 96-well microplate reader (MRX microplate reader, Dynatech Laboratories, Denkendorf, USA). Experiments were performed in four to eight biological replicates and each experiment was carried out six times.

3. Results

3.1. Casting of silver-alloys

The alloys were produced along the routine described before. The main difference regarding the production of both alloys concerns the extending of the content of the alloying element calcium. For casting the AgCa alloy, the amount of calcium had to be extended 20% to achieve a satisfying content of calcium in the ingot. In contrast to the AgCa alloy, an extended calcium amount was not necessary for the AgCaLa alloying system to achieve satisfying contents of calcium.

3.2. Microstructural characterization

3.2.1. Scanning electron microscopy (SEM)

The microstructure of the AgCa alloy is depicted in Fig. 1(a) and (b). The alloy shows coarse lamellae and a mixture of at least two phases between the lamellae. The interlamellar structure is finer compared to coarse lamellae. The light microscopy images suggest that, besides the fine structured phase mixture, a third phase with a bright appearance exists preferably along the lamellae. The microstructure of the AgCaLa alloy is depicted in Fig. 1(c) and (d). The ternary AgCaLa alloy contains at least three phases. One phase with a bright appearance surrounded by a relatively fine lamellae structure filled with a third phase (Fig. 1c,d).

The elevations and cracks that are partially apparent are probably caused by corrosion during preparation.

3.2.2. Energy dispersive spectroscopy (EDS)

As described, there are minor changes in chemical composition due to evaporation and oxidization during the casting of the alloys. Additionally, segregations could lead to varying content of alloying elements in different parts of the ingot. To determine this, the chemical composition was measured by EDS. Table 1 summarizes the chemical composition of the selected ingots. To minimize the influence of the selected area, the content was determined as the average of the measurement in different positions of the ingot.

The EDS-mappings given in Fig. 2 enable the identification of the composition and structure of the different phases in the same sections as in Fig. 1(a) and (d). Fig. 2(a) and (b) shows the contribution of silver and calcium for the binary alloy AgCa. The lamellae structures proofs a homogenous distribution of calcium for 14 wt.-% Ca. The identification of the content of the single phases is affected by inaccuracies due to the measuring method e.g. the effect of neighboring or underlying phases. Accordingly, the identification of the composition of the finely structured phases is not possible with EDS. The mixture of both phases has an average content of 19 wt.-% calcium. In the lower part of Fig. 2(a) and (b) some areas with significantly low contents of calcium, having only 2 wt.-% calcium, along the lamellae, are depicted. In summary, the EDS-mapping indicates the formation of three to four different phases. Whether the phase with low content of calcium occurring along the lamellae is identical to one of the finely structured phases between the lamellae could not be identified clearly according to the EDS-mappings.

For the AgCaLa alloy, the formation of at least three different phases is detected (Fig. 2c–e). The phase with a bright appearance identified at the SEM-image (Fig. 1c,d) is enriched with lanthanum and contains 17 wt.-% lanthanum and only 6 wt.-% calcium.

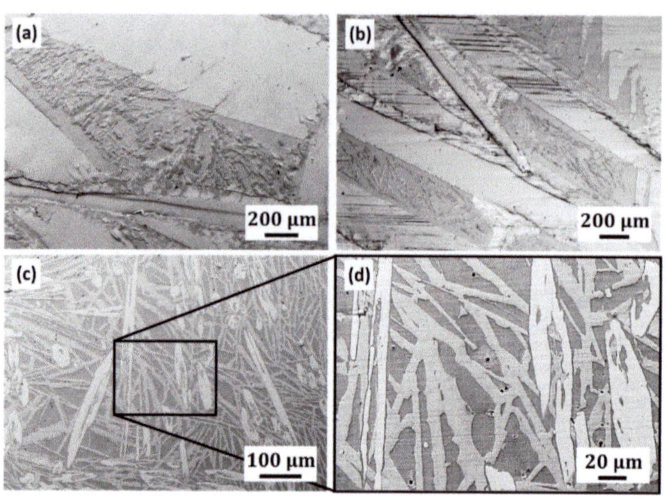

Fig. 1. SEM-Secondary Electron (SEM-SE) (a, c, d) and light microscopy (b) - cross-sections showing the microstructure of (a, b) AgCa and (c, d) AgCaLa alloys.

Table 1
Average composition of the alloys measured by EDS.

Alloy	Composition (wt%)		
	Ag	Ca	La
AgCa	88.4	11.6	–
AgCaLa	82.8	12.1	5.1

Contrariwise the surrounding lamellae contain less calcium and lanthanum (3 wt.-% lanthanum; 12 wt.-% calcium). The phase between the lamellae is enriched in calcium and contains 17 wt.-% and only 4 wt.-% lanthanum. Therefore, the EDS-mapping indicates three different phases, one with high calcium content in combination with low lanthanum content, one phase the other way round, and the third one with high silver content.

3.2.3. X-ray diffraction

The phases in both alloys were investigated by XRD scans to identify the existing phases, which are presented in Fig. 3. The intensity of the peaks is low due to coarse grains. For the AgCa alloy, the best-fitting two phases Ag_2Ca (I mma) and AgCa (C mcm) have both an orthorhombic crystal structure and are marked in Fig. 3. Analogous to the binary system, the Ca-rich, and La-rich phases of the AgCaLa alloy were identified to have an orthorhombic crystal structure. The peaks from the Ca-rich and La-rich phases are marked appropriately in Fig. 3. Nevertheless, an overlap of the peaks from different phases is observed too.

3.2.4. Cooling curves

The cooling curves of AgCa and AgCaLa alloys are superimposed in Fig. 4. To identify the alloy-dependent features, the cooling curves are shifted on the timeline. For the interpretation of the cooling curves, the ambient conditions during solidification and cooling have to be taken into account. The ingots are preheated to 700 °C and taken out of the oven shortly before casting. After casting the heat is transferred to the environment with a constant temperature of 20 °C. Consequently, the temperature difference decreases with decreasing temperature, and the cooling rate decreases due to this mechanism.

Hence, the globally decreasing cooling rate is not referred to as a phase transformation and has not to be taken into account.

Apart from this, the cooling curves of both alloys show inflection points and stopping points indicating phase transformations. The cooling curve of AgCa shows two stopping respectively inflection points (Fig. 4). The first point is observed at 685 °C and linked with only a shortly stopping of the decreasing temperature, hence, identifying a phase transformation. The following decreasing temperature is decelerated which indicates a continuous phase transformation. The decreased cooling in the region between 685 °C and 594 °C becomes clear in comparison to the faster cooling below 594 °C. At a temperature of 594 °C, a second stopping point with a more distinct temperature plateau suggests the end of the continuous phase transformation and indicates a new phase transformation. Consequently, it can be considered, that three phase transformations are identified according to the cooling curve, two at certain temperatures, and one continuously between.

The cooling curve of the AgCaLa alloy shows significant points in the same temperature zone as the AgCa alloy. Starting with a stopping point with supercooling at 694 °C indicating a phase transformation. At 670 °C, another stopping point with only a small hold-up is detected. The following cooling is decelerated suggesting continuous phase transformation. At 605 °C, a stopping point with distinct supercooling followed by slow cooling can be seen. The reduced cooling rate until 590 °C indicates one phase transformation with imprecise characteristics in the cooling curve, or one phase transformation followed by a continuous transformation. In summary, the cooling curve for AgCaLa indicates more phase transformations than the cooling curve of the binary alloy. At least three transformations at a certain temperature and one or two continuous transformations are suggested.

3.3. Corrosion measurements

3.3.1. Open circuit potential

The OCP curves for AgCa and AgCaLa, as well as XIP-1000 TWIP steel (22 wt.-% Mn, 0,6 wt.-% C, Fe balance), are summarized in Fig. 5. All measurements were performed in the equilibrium condition. The

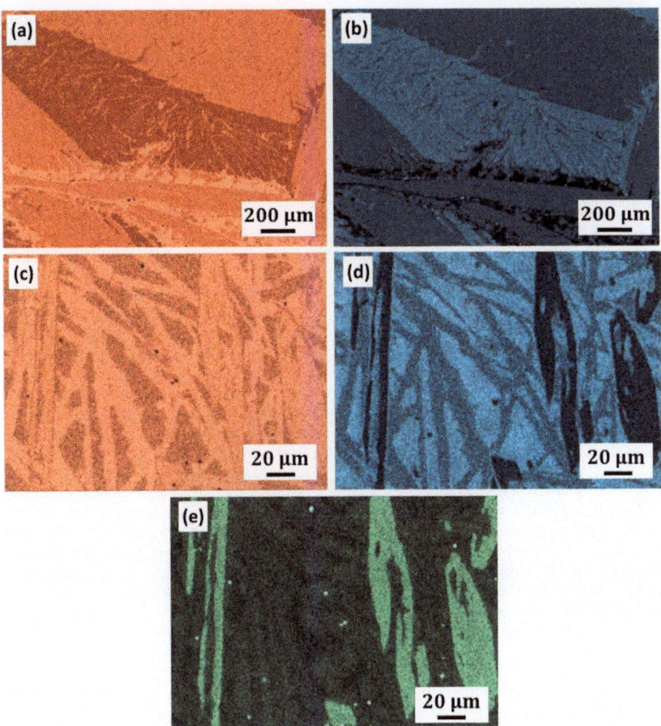

Fig. 2. EDS-Mappings presenting the distribution of (a, c) Ag, (b, d) Ca and (e) La in the (a, b) AgCa and (c, d, e) AgCaLa alloys, corresponding SEM-SE: Fig. 1(a, d).

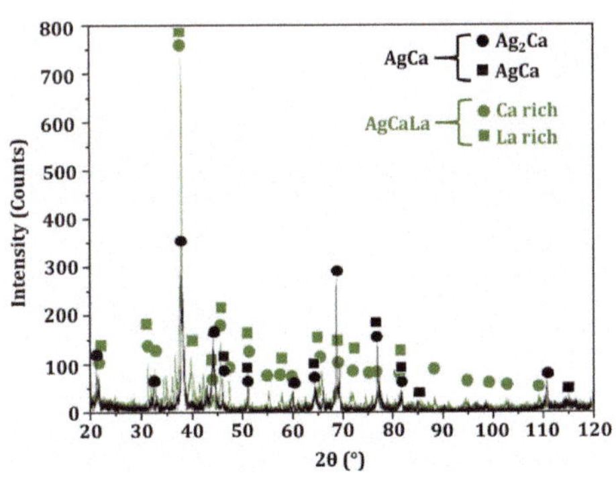

Fig. 3. X-ray diffraction patterns of AgCa and AgCaLa alloys.

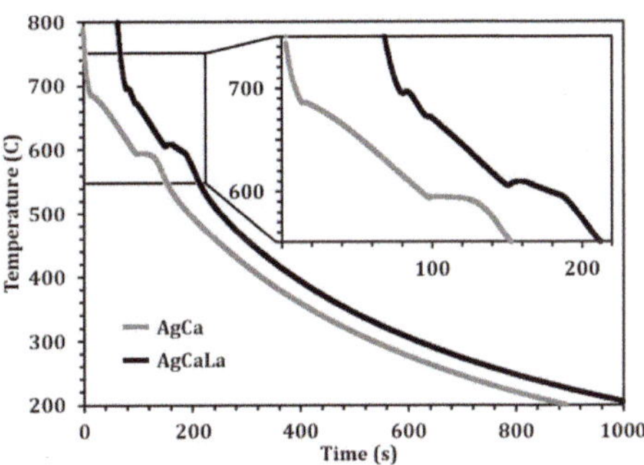

Fig. 4. Cooling curves for the AgCa and AgCaLa alloys.

drifting of the curves indicates the unstable states during the measurement. One reason could be that the AgCaLa alloy reacts with some constituents of Ringer's Lactate solution to form a protective layer. This layer increases the OCP. However, the potential curves move towards a final value at the end of the measurement. Due to the used electrolyte, the conditions are similar to the conditions in the human organism, consequently the reactions and developed surfaces are similar. Additional OCP measurement of XIP-1000 TWIP steel, which is selected as a possible matrix-material; showed the OCP to be about −500 mV in the same conditions (Table 2). As the measurements were all carried out under similar conditions comparability is given.

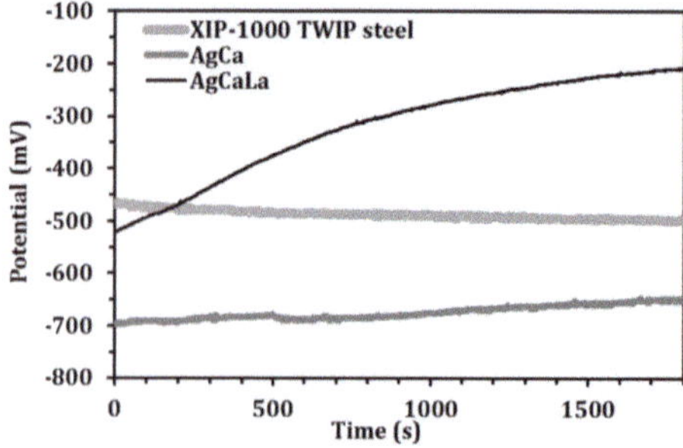

Fig. 5. Results of the OCP measurements for AgCa, AgCaLa alloys, and XIP-1000 TWIP steel in Ringer's Lactate solution (results are referred to the SHE).

Table 2
Average open circuit potential of all alloys in Ringer's Lactate solution vs. SHE.

Alloy	OCP (mV)
AgCa	−650
AgCaLa	−210
XIP-1000 TWIP steel	−500

3.3.2. Immersion test

This section presents the results of a 2000 h immersion (eleven weeks) of AgCa and AgCaLa alloys in Ringer's Lactate solution to trace the degradation behavior over a long period (Fig. 6). During the experiments, the mass of the AgCa and AgCaLa alloys were observed to decrease overall with time. However, the AgCa alloy shows erratic degradation. After a significant initial mass loss during the first 24 h, the weight remains nearly constant and shows only slow but continuous mass loss within the next 265 h. Afterward, the mass loss increases again distinctly between 265 h and 360 h. From that point on until the end of the immersion test, the mass decreases discontinuously. The visual observation of the samples reveals striking differences at these time steps. After the first day, the previous blank metallic surface becomes matt homogenous and the structure is slightly visible. Later (at 360 h), the sample shows cracks along with the microstructure and inhomogenous deposits. As the immersion continues, the separation in the microstructure becomes more significant, but no decay of the sample is observed until the end of the test. Simultaneously the thickness of the evolving corrosion products seems to increase, some corrosion products are detached and blank metallic surface areas originate.

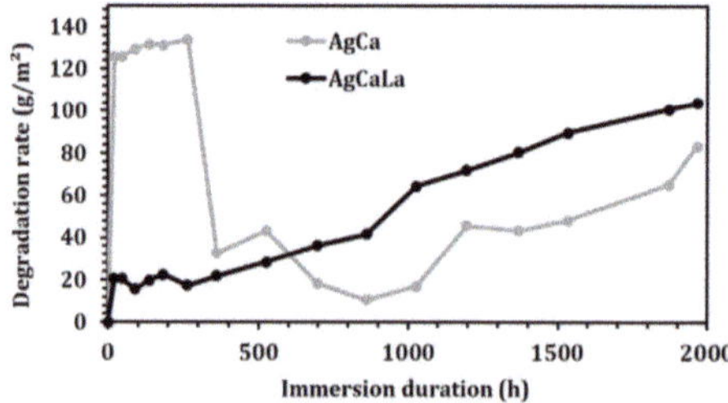

Fig. 6. Degradation rates of AgCa and AgCaLa alloys in Ringer's Lactate solution for 2000 h.

Compared to this, the AgCaLa alloy shows an almost continuous mass decrease after the initial significant mass reduction. The surface of the samples shows a consistent behavior. After 24 h, a homogenous matt surface developed with some remaining metallic blank areas which disappear until 180 h. Up to the end of the immersion test, the surface does not change significantly, minor depositions can be observed and the surface roughness seems to increase.

The measurement of the pH-values shows a slight increase in the value compared to the new Ringer's Lactate solution with a determined pH-value of 8. The pH-values of both alloys show a similar trend. Until the first 400 h of immersion, the measured value is approximately 8.5 and at the end of the immersion test, the value is 8.2.

The investigation of degraded samples shows significant differences between the alloys (Fig. 7). The AgCa alloy has an inhomogenous degradation of different phases (Fig. 7b). The lamellae show typically less degradation products whereas the phases between the lamellae exhibit distinct deposits. Additionally, a decay along some of the lamellae takes place. In contrast, the AgCaLa alloy is coated by a continuous layer which makes a differentiation between the degradation of the different phases impossible. Consequently, it can be stated that the degradation of AgCaLa (Fig. 7c,d) alloy is more homogenous regarding surface and time.

Furthermore, the corrosion products of both alloys show differences. The products on the surface of AgCa alloy are rough and seem to be similar structures grown beside each other, while the AgCaLa alloy shows more connected deposits with round elevations.

3.4. Cytocompatibility tests

In our experiments, a 24-hour incubation of the cells with the degradation media of the AgCaLa alloy has no negative influence on cell viability and cell proliferation (Fig. 8). At nearly each degradation volume and each dilution, the viability and proliferation capacity of the cells is above the limit of 70% of viability or proliferation, respectively. An exception was seen in the proliferation of the HOS cells that were incubated with undiluted degradation media. In this group, the values were below the limit of 70%. However, incubation with these degradation media showed no reduction in viability.

4. Discussion

4.1. Microstructures and phase identification

The determination of the developed phases requires a connection of the results of all examinations and available information for the binary phase diagrams of the AgCa and AgLa alloying systems (Fig. 9). Regarding the ternary system AgCaLa, no information is on hand. Furthermore, it has to be taken into account, that the phase diagrams describe the equilibrium state and the solidification during the presented examinations probably appears not in equilibrium due to fast cooling.

A hint of non-equilibrium solidification gives the cooling curve of AgCa (Fig. 4) in combination with the AgCa phase diagram (Fig. 9a) as the identified stopping points are in accordance with the formation-temperature of Ag_9Ca_2 at 667 °C and Ag_2Ca at 597 °C. These phases do not exist simultaneously in the equilibrium state as the transformations take place at different concentrations above and under 10 wt.-% calcium. Accordingly, segregation takes place and phases with high and low calcium content are formed simultaneously. The existence of Ag_2Ca is confirmed by the XRD. As the measured calcium content of the lamellae is 14 wt.-% it can be considered, that the lamellae consist of Ag_2Ca with a nominal content of approximately 15.5 wt.-%. The finely structured phases between the lamellae can be allocated as the eutectic mixture of Ag_2Ca

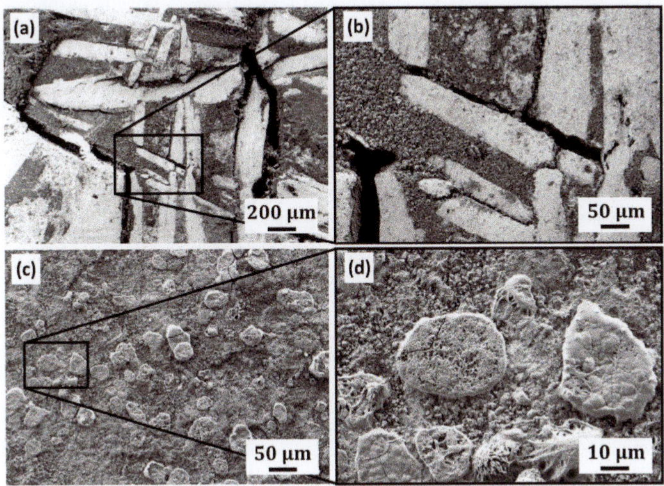

Fig. 7. Scanning electron microscopy of the sample surfaces after the immersion test in Ringer's Lactate solution of (a, b) AgCa, and (c, d) AgCaLa alloy.

and AgCa as the measured content of 19 wt.-% fits the eutectic point at 19.5 wt.-%. Although the eutectic reaction is not detectable in the cooling curve, the formation of eutectic is conclusive due to the proven existence of AgCa by XRD, the measured calcium content,

and the eutectic like structure in the cross-sections. Furthermore, Alexander et al. [43] reported that the temperatures of eutectic and solid-state transformations are partially poorly reproducible. They stated that calcium shows sensitivity to impurities and the applied

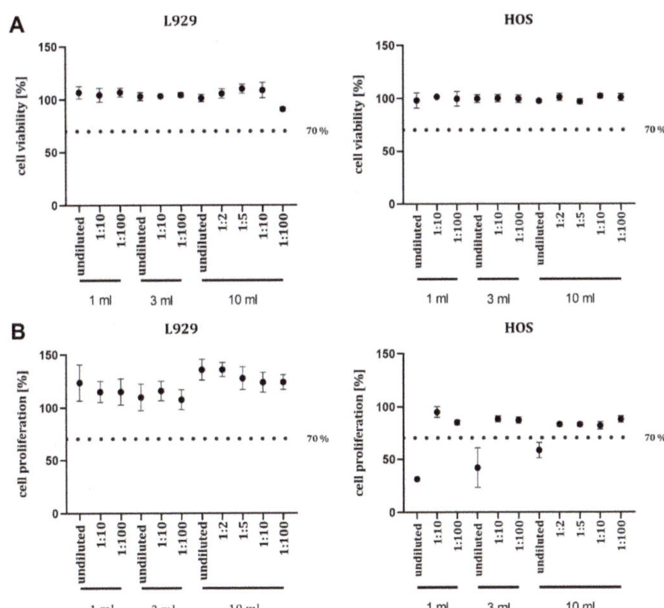

Fig. 8. Effects of incubation with degradation media on HOS and L929 cells after 24 h; results are shown for cell viability (A) and cell proliferation (B) in % referred to the control; the dotted line represents the lower limit at 70% to reduced cell viability and cell proliferation. mean ± SD, n = 6.

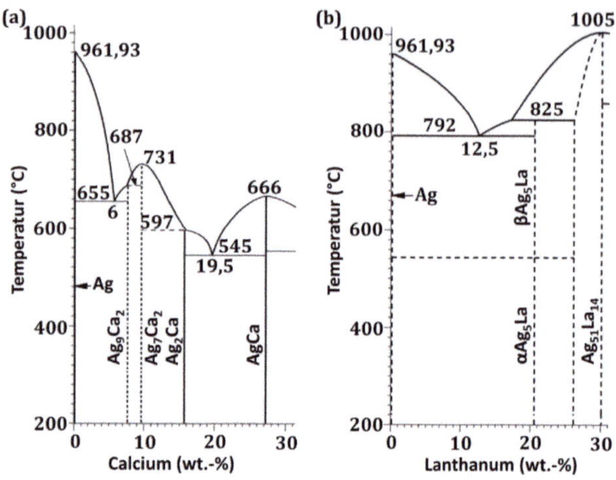

Fig. 9. Relevant sections of the phase diagrams of the (a) Ag-Ca- and (b) Ag-La-systems, modified according to [34,44,45].

calcium has a purity of merely 98.5 wt.-%. Thus, aberrations from phase diagrams are explainable.

As all declared phases contain more calcium than the overall measured value, the existence of phases with a low calcium content is implied. The identification of the low content phases is not as clear since the cooling curves suggest the formation of Ag$_9$Ca$_2$ with a calcium content of 8 wt.-%. Besides, for the phases with low calcium content along the lamellae a content of 2 wt.-% was determined via EDS. Due to the influence of neighboring phases during EDS, it is possible, that the phases with low calcium content are almost pure silver.

Concerning the phases determined before, the solidification process is discussed as follows: Due to the phase diagram, the solidification starts with the formation of Ag$_7$Ca$_2$ which is ongoing continuously until this phase is mainly transformed to Ag$_2$Ca with a lamellae structure. The segregation causes the simultaneous generation of phases with low calcium content containing some Ag$_9$Ca$_2$ according to the cooling curve, as well as pure silver due to EDS analysis. Finally, the eutectic of Ag$_2$Ca and AgCa is formed. Consequently, the main phases of the examined AgCa alloy are Ag$_2$Ca as part of the eutectic and the lamellae, AgCa as part of the eutectic, and phases with low silver content along the lamellae.

A certain identification of phases of the AgCaLa alloy is not possible due to a lack of information. The optical appearance suggests that the lanthanum-rich phase is surrounded by the other and consequently solidify first. This is conclusive as the binary system AgLa shows solidification at a higher temperature level than the AgCa system (Fig. 9a,b). Probably the solidification is ongoing with the lamellae structure and completed with the areas between lamellae. The resulting microstructure shows similarity to the lamellae structure of AgCa, but the lamellae structure of AgCaLa is much finer and interrupted by the lanthanum rich phases. However, even the identified inflection and stopping points, the crystal structure determined by XRD, and the chemical composition of different phases show similarities to binary alloys. It is striking, that the content of lanthanum and calcium in sum corresponds well to the content of phases of binary alloys. The sum of lanthanum and calcium of the lamellae phase is 15 wt.-% and 21 wt.-% for the phase between the lamellae corresponding to the content of Ag$_2$Ca and the eutectic of Ag$_2$Ca and AgCa in the binary AgCa-system. Thus, some of

the calcium seems to be replaced by lanthanum. The same relation can be observed for the lanthanum rich phases as they contain in sum 23 wt.-% alloying elements corresponding to the phase Ag$_5$La which contains approximately 20.5 wt.-% lanthanum (Fig. 9b). Consequently, the theory postulated in the introduction, that phases similar to the phases in binary alloys are formed, is confirmed by the microstructural investigations.

The cooling curve of AgCaLa shows conspicuities at a similar temperature level as the AgCa alloy confirming the formation of phases with similar properties in both examined alloys. The contribution between observed phase transformation in the cooling curve and the microstructural investigation is not possible as the cooling curve suggests at least three transformations at certain and one continuous transformation, but only three different phases are identified via cross-sections. Furthermore, the phase with high lanthanum content eventually develops at elevated temperatures according to the binary AgLa phase diagram (Fig. 9b). The temperature decreasing in the possible temperature area of consolidation of phases with high lanthanum content is very fast and consequently, a significance is eventually not noticeable. Also possible is, that the first stopping point with supercooling contributes to the formation of lanthanum-rich phases. Hence, the next small stopping point results from the formation of phase-related with Ag$_7$Ca referring to a continuous transformation. Then the large supercooling would contribute to the transformation of this phase to Ag$_2$Ca. Finally, the consolidation of the phase between the lamellae could generate the following decreased cooling rate. Nevertheless, the phases between the lamellae show no detectable eutectic structure and are homogenous. Therefore, the similarity of this phase regarding the mixture of phases between the lamellae of the AgCa alloy is questionable. However, it is possible, that some not detectable phases exist in small amounts or a transformation takes place without a noticeable impact on the cooling curve.

But in summary, it can be stated that the examined AgCaLa alloy consists mainly of three phases: One lanthanum-rich, one silver-rich with lamellae shape, and one calcium-rich phase. These three phases show similarities to the phase Ag$_5$La of the AgLa-system, the phase Ag$_2$Ca, and the eutectic of Ag$_2$Ca and AgCa of the AgCa-system. The microstructure of the AgCaLa alloy is finer compared to the structure of AgCa.

4.2. Degradation response

Pure silver does not show much degradation [33]. However, AgCa alloys exhibit a faster degradation leading to a higher mass-loss rate [33].

The measured OCP is of interest regarding two different aspects. First, the OCP gives hints concerning the degradation behavior, as a low potential indicates faster degradation. Nevertheless, other effects are possible e.g. due to potential differences between different phases. Second, the OCP determines how effective the alloy can increase the degradation of iron-based alloys. As the concept of FeAg alloys is based on the potential difference, degradable alloys with high electrochemical potential are required. Hence, there is a prerequisite for silver alloys with high OCP and adequate degradation behavior. Higher contents of less noble alloying elements should result in a reduction of the electrochemical potential and consequently in increased degradation. The OCPs of AgCa and AgCaLa alloys are in between the OCP of pure silver, pure calcium, and pure lanthanum. Nonetheless, the OCP of the AgCaLa alloy is higher than the one of the AgCa alloy; although more elements with low potential are added. The increased OCP may be attributed to the formation of intermetallic phases with relatively high potentials. The resulting potential of the AgCaLa alloy (−210 mV) is nobler than the potential of the iron-based alloy XIP-1000 (−500 mV). Consequently, the AgCaLa alloy is suitable for the combination with XIP-1000 regarding the OCP contrary to the AgCa alloy with a more negative potential (−650 mV).

The electrochemical potential of the different phases of the alloys is not known, but regarding the AgCa alloy, it can be assumed that the potential of AgCa is lower than the potential of Ag_2Ca due to the faster degradation of the AgCa containing eutectic. Further, it can be stated, that the decay of the sample results from the high potential difference between the phases with very low calcium content along the lamellae and the surrounding phases, especially the eutectic with AgCa. Accordingly, the potential difference between the phases leads to increased degradation. Unfortunately, the increased degradation between the phases results in the decay of the sample due to the coarse microstructure of the AgCa alloy. Although no decay of samples is observed during the test period, the evolution of the samples indicates soon decay. The resulting finer particles with sharp edges are undesired as they can cause complications and further dispersion in the human body. Furthermore, the degradation of the AgCa alloy is not continuous concerning mass loss and surface evolution. The initial mass loss during the first 24 h and the following almost constant weight result from the formation of a degradation inhibiting layer on the surface. At the beginning of immersion, the blank metallic surface is in contact with the immersion medium. Hence, direct exchange between surface and electrolyte is possible, resulting in pronounced degradation accompanied by the layer formation. The almost constant weight after 24 h might result from this layer. But between 265 h and 360 h, this barrier is penetrated corresponding to the observed formation of cracks. Due to the reoccurring exchange, further degradation takes place and the mass of the corrosion products increases related to the increased weight of the samples. The further irregular course of sample weight can be explained by statistical formation and detachment of corrosion products. Comprehensively, it can be stated that the unpredictable degradation of AgCa is caused by inhomogeneous layer formation and is not adapted to the intended application.

However, the AgCaLa alloy shows more homogenous degradation in the desired manner probably due to the finer microstructure. The mass loss during the first 24 h is also distinctive but significantly less than for AgCa. Considering the development of the sample surface, this is also associated with the deposition of degradation products. Until 180 h after the beginning of the immersion test, these degradation products form a continuous layer. The layer formation is referred to as nearly constant sample weight, as the degradation proceeds and simultaneously more degradation products are generated. Afterward, the weight decrease proceeds continuously in the required manner. So it can be stated, that the layer containing the degradation products on AgCaLa alloy does not inhibit the exchange, but minimizes it.

For better comparability, the measured degradation rate is extrapolated to a year based on the final measurement after eleven weeks and referred to the density to achieve a value for the reduction of the thickness. Due to inhomogenous degradation, this value reflects not real behavior, but especially for the more homogenous degradation of AgCaLa, it is a good approach. Furthermore, degradation proceeds continuously, in particular for the more relevant AgCaLa alloy. Thus it can be stated, that the formation of corrosion products is overall not associated with the formation of a protective layer. As a result, the AgCa alloy loses 60 µm/a and the AgCaLa alloy 80 µm/a.

During the degradation of FeAg alloys, the silver alloys would probably remain nearly unmodified due to their more noble character until the surrounding matrix is degraded. Subsequently, their dissolution will start. As the silver phases originated from additive manufacturing have particle sizes of approximately 20–50 µm [18], the particles would dissolve during only a few months due to the corrosion attack from all sides. The nearly constant pH-values from 400 h to the end of the immersion test confirm that a constant continuing degradation takes place.

Cytotoxicity examinations show promising results regarding the influence of the degradation media on cell viability and proliferation confirming the biocompatibility, as nearly all indicators are significantly better than the critical limit. Only undiluted degradation media result in a diminished proliferation rate in HOS cells. Although HOS cells proliferate in a reduced manner they show high viability rates. Thus, the AgCaLa alloy is a very promising candidate for an iron-based degradable implant with adapted degradation behavior and good biocompatibility. The AgCaLa alloy combines promising biocompatibility with suitable degradation behavior, high electrochemical potential, and well processability. There are several reasons for the addition of lanthanum to an AgCa alloy: First, the AgCaLa alloys investigated exhibit a steady monotonic increasing degradation. Second, AgCaLa alloys can be cast easily. Third, AgCaLa shows a more homogenous degradation regarding time and surface and is thus favorable for the intended application. Here, the addition of lanthanum leads to a significant increase in the OCP.

5. Conclusions

The present investigation highlights the potential of AgCaLa alloys as additive alloying material for iron-based bioresorbable implants. There are several reasons for the addition of lanthanum to the AgCa-alloy in the intended application:

1. Lanthanum shifts the OCP to more positive values (Fig. 5),
2. Lanthanum increases the degradation rate (Fig. 6),
3. Lanthanum leads to a more homogenous degradation (Fig. 7),
4. Lanthanum enhances the processability in terms of melting and casting,
5. Lanthanum does not affect cell viability and proliferation (Fig. 8).

In this regard, the attained results reveal that AgCaLa alloys are promising for the addressed application. So, the AgCaLa alloy can be recommended as the desired alloying system for iron-based bioresorbable implants with an adapted degradation rate due to galvanic coupling. Moreover, only small quantities of lanthanum are required to achieve significant changes in the microstructure as well as in the degradation rate of silver alloys. As expected, the resulting microstructure and properties of AgCaLa are a good combination of the binary alloys due to the immiscibility of calcium and lanthanum.

For further investigations, these ternary alloys will be cast in a higher quantity for gas atomization. The gas atomized AgCaLa alloy will be mechanically mixed with gas atomized XIP-1000 TWIP steel powder and processed via laser beam melting to investigate the interaction of the AgCaLa alloy with TWIP steel and the resulting degradation behavior.

CRediT authorship contribution statement

Jan Krüger: Writing - review & editing, Methodology, Investigation. **Kay-Peter Hoyer:** Writing - review & editing, Project administration. **Viviane Filor:** Writing - review & editing, Methodology, Investigation. **Sudipta Pramanik:** Writing - original draft. **Manfred Kietzmann:** Project administration, Funding acquisition. **Jessica Meißner:** Writing - review & editing. **Mirko Schaper:** Project administration, Funding acquisition.

Declaration of Competing Interest

The authors declare that they have no conflict of interest.

Acknowledgment

The authors gratefully thank the German Research Foundation (DFG) for financial support. This work results from the projects (SCHA1484/44-1 and KI361/8-1). Furthermore, the authors also would like to thank Sabrina Beumer, Thomas Janzen, Dr. Barbara Flöing-Hering, and Mathias von Spalden for promoting the production and examination of the silver alloys.

References

[1] D. Carluccio, A.G. Demir, M.J. Bermingham, M.S. Dargusch, Challenges and opportunities in the selective laser melting of biodegradable metals for load-bearing bone scaffold applications, Metall. Mater. Trans. A 51 (2020) 3311–3334, https://doi.org/10.1007/s11661-020-05829-7

[2] M. Peuster, C. Hesse, T. Schloo, C. Fink, P. Beerbaum, C. von Schnakenburg, Long-term biocompatibility of a corrodible peripheral iron stent in the porcine descending aorta, Biomaterials 27 (28) (2006) 4955–4962, https://doi.org/10.1016/j.biomaterials.2006.05.029

[3] H. Hermawan, M. Moravej, D. Dubé, M. Fiset, D. Mantovani, Degradation behaviour of metallic biomaterials for degradable stents, Adv. Mater. Res. 15–17 (2007) 113–118, https://www.scientific.net/AMR.15-17.113

[4] H. Hermawan, D. Dubé, D. Mantovani, Development of degradable Fe-35Mn alloy for biomedical application, Adv. Mater. Res. 15–17 (2007) 107–112, https://doi.org/10.4028/www.scientific.net/AMR.15-17.107

[5] N. Abbaspour, R. Hurrell, R. Kelishadi, Review on iron and its importance for human health, J. Res. Med. Sci. 19 (2) (2014) 164–174.

[6] M. Peuster, P. Wohlsein, M. Brügmann, M. Ehlerding, K. Seidler, C. Fink, H. Brauer, A. Fischer, G. Hausdorf, A novel approach to temporary stenting: degradable cardiovascular stents produced from corrodible metal - results 6-18 months after implantation into New Zealand white rabbits, Heart 86 (2001) 563–596 https://dx.doi.org/10.1136%2Fheart.86.5.563.

[7] J. Kuhlmann, I. Bartsch, E. Willbold, S. Schuchardt, O. Holz, N. Hort, D. Höche, W.R. Heineman, F. Witte, Fast escape of hydrogen from gas cavities around corroding magnesium implants, Acta Biomater. 9 (10) (2013) 8714–8721, https://doi.org/10.1016/j.actbio.2012.10.008

[8] J. Čapek, Š. Msallamová, E. Jablonská, J. Lipov, D. Vojtěch, A novel high-strength and highly corrosive biodegradable Fe-Pd alloy: structural, mechanical and in vitro corrosion and cytotoxicity study, Mater. Sci. Eng. C 79 (2017) 550–562, https://doi.org/10.1016/j.msec.2017.05.100

[9] M. Schinhammer, C.M. Pecnik, F. Rechberger, A.C. Hänzi, J.F. Löffler, P.J. Uggowitzer, Recrystallization behavior, microstructure evolution and mechanical properties of biodegradable Fe–Mn–C(–Pd) TWIP alloys, Acta Mater. 60 (6–7) (2012) 2746–2756, https://doi.org/10.1016/j.actamat.2012.01.041

[10] W. Lin, G. Zhang, P. Cao, D. Zhang, Y. Zheng, R. Wu, L. Qin, G. Wang, T. Wen, Cytotoxicity and its test methodology for a bioabsorbable nitrided iron stent, J. Biomed. Mater. Res. Part B Appl. Biomater. 103 (4) (2014) 764–776, https://doi.org/10.1002/jbm.b.33246

[11] J.D. Bobyn, E.S. Mortimer, A.H. Glassman, C.A. Engh, J.E. Miller, C.E. Brooks, Producing and avoiding stress shielding. laboratory and clinical observations of noncemented total hip arthroplasty, Clin. Orthop. Relat. Res. 274 (1992) 79–96, https://doi.org/10.1097/00003086-199201000-00010

[12] M. Schinhammer, A.C. Hänzi, J.F. Löffler, P.J. Uggowitzer, Design strategy for biodegradable Fe-based alloys for medical applications, Acta Biomater. 6 (5) (2010) 1705–1713, https://doi.org/10.1016/j.actbio.2009.07.039

[13] H. Hermawan, A. Purnama, D. Dube, J. Couet, D. Mantovani, Fe–Mn alloys for metallic biodegradable stents: degradation and cell viability studies, Acta Biomater. 6 (5) (2010) 1852–1860, https://doi.org/10.1016/j.actbio.2009.11.025

[14] M.S. Dargusch, A.D. Manshadi, M. Shahbazi, J. Venezuela, X. Tran, J. Song, N. Liu, C. Xu, Q. Ye, C. Wen, Exploring the role of manganese on the microstructure, mechanical properties, biodegradability, and biocompatibility of porous iron-based scaffolds, ACS Biomater. Sci. Eng. 5 (4) (2019) 1686–1702, https://doi.org/10.1021/acsbiomaterials.8b01497

[15] J. Hufenbach, J. Sander, F. Kochta, S. Pilz, A. Voss, U. Kühn, A. Gebert, Effect of selective laser melting on microstructure, mechanical, and corrosion properties of biodegradable FeMnCS for implant applications, Adv. Eng. Mater. 22 (2020) 2000182, https://doi.org/10.1002/adem.202000182

[16] H. Hermawan, D. Dube, D. Mantovani, Degradable metallic biomaterials: design and development of Fe–Mn alloys for stents, J. Biomed. Mater. Res. Part A 93 (1) (2010) 1–11, https://doi.org/10.1002/jbm.a.32224

[17] M. Schinhammer, P. Steiger, F. Moszner, J.F. Löffler, P.J. Uggowitzer, Degradation performance of biodegradable FeMnC(Pd) alloys, Mater. Sci. Eng.: C 33 (4) (2013) 1882–1893, https://doi.org/10.1016/j.msec.2012.10.013

[18] T. Niendorf, F. Brenne, P. Hoyer, D. Schwarze, M. Schaper, R. Grothe, M. Wiesener, G. Grundmeier, H.J. Maier, Processing of new materials by additive manufacturing: iron-based alloys containing silver for biomedical applications, Metall. Mater. Trans. A 46 (2015) 2829–2833, https://doi.org/10.1007/s11661-015-2932-2

[19] T. Huang, J. Cheng, Y.F. Zheng, In vitro degradation and biocompatibility of Fe–Pd and Fe–Pt composites fabricated by spark plasma sintering, Mater. Sci. Eng.: C 35 (2014) 43–53, https://doi.org/10.1016/j.msec.2013.10.023

[20] T. Huang, J. Cheng, D. Bian, Y. Zheng, Fe-Au and Fe-Ag composites as candidates for biodegradable stent materials, J. Biomed. Mater. Res. Part B Appl. Biomater. 104 (2016) 225–240, https://doi.org/10.1002/jbm.b.33389

[21] T.B. Karchmer, E.T. Giannetta, C.A. Muto, B.A. Strain, B.M. Farr, A randomized crossover study of silver-coated urinary catheters in hospitalized patients, Arch. Intern. Med. 160 (2000) 3294–3298, https://doi.org/10.1001/archinte.160.21.3294

[22] M. Wiesener, K. Peters, A. Taube, A. Keller, K.-P. Hoyer, T. Niendorf, G. Grundmeier, Corrosion properties of bioresorbable FeMn-Ag alloys prepared by selective laser melting, Mater. Corros. 68 (2017) 1028–1036, https://doi.org/10.1002/maco.201709478

[23] P.S. Bagha, M. Khakbiz, S. Sheibani, H. Hermawan, Design and characterization of nano and bimodal structured biodegradable Fe-Mn-Ag alloy with accelerated corrosion rate, J. Alloy. Compd. 767 (2018) 955–965, https://doi.org/10.1016/j.jallcom.2018.07.206

[24] L.E. Murr, E. Martinez, K.N. Amato, S.M. Gaytan, J. Hernandez, D.A. Ramirez, P.W. Shindo, F. Medina, R.B. Wicker, Fabrication of metal and alloy components by additive manufacturing: examples of 3D materials science, J. Mater. Res. Sci. 1 (1) (2012) 42–54, https://doi.org/10.1016/S2238-7854(12)70009-1

[25] C. Yan, L. Hao, A. Hussein, D. Raymont, Evaluations of cellular lattice structures manufactured using selective laser melting, Int. J. Mach. Tools Manuf. 62 (2012) 32–38, https://doi.org/10.1016/j.ijmachtools.2012.06.002

[26] A.B.G. Lansdown, A. Williams, How safe is silver in wound care? J. Wound Care 13 (4) (2004) 131–136, https://doi.org/10.12968/jowc.2004.13.4.26596

[27] A. Wadhera, M. Fung, Systemic argyria associated with ingestion of colloidal silver, Dermatol. Online J. 11 (1) (2005).

[28] M.A. Hollinger, Toxicological aspects of topical silver pharmaceuticals, Crit. Rev. Toxicol. 26 (3) (1996) 255–260, https://doi.org/10.3109/10408449609012524

[29] World Health Organization, Silver in drinking-water background document for development of WHO guidelines for drinking-water quality guidelines for drinking-water quality, second ed., Health Criteria and other Supporting Information 2 World Health Organization, Geneva, 1996.

[30] K. Mijnendonckx, N.L.J. Mahillon, S. Silver, R. Van Houdt, Antimicrobial silver: uses, toxicity and potential for resistance, Biometals 26 (2013) 609–621, https://doi.org/10.1007/s10534-013-9645-z

[31] B. Shahare, M. Yashpal, G. Singh, Toxic effects of repeated oral exposure of silver nanoparticles on small intestine mucosa of mice, Toxicol. Mech. Methods 23 (3) (2013) 161–167, https://doi.org/10.3109/15376516.2013.764950

[32] J. Tang, L. Xiong, S. Wang, J. Wang, L. Liu, J. Li, F. Yuan, T. Xi, Distribution, translocation and accumulation of silver nanoparticles in rats, J. Nanosci. Nanotechnol. 9 (2009) 4924–4932, https://doi.org/10.1166/jnn.2009.1269

[33] K.-P. Hoyer, M. Schaper, Alloy Design for Biomedical Applications in Additive Manufacturing, TMS 2019 148th Annual Meeting & Exhibition Supplemental Proceedings, Springer International Publishing, 2019, pp. 475–484, https://doi.org/10.1007/978-3-030-05861-6_44

[34] ASM Alloy P D D ASM Alloy Phase Diagram Database. (1999).

[35] European Food Safety Authority, Overview on Tolerable Upper Intake Levels as derived by the Scientific Committee on Food (SCF) and the EFSA Panel on Dietetic Products, Nutrition and Allergies (NDA), Summary of Tolerable Upper Intake Levels – version 4, 2018. (https://www.efsa.europa.eu/sites/default/files/assets/UL_Summary_tables.pdf) (Accessed 29.09.2020).

[36] G. Pagano, M. Guida, F. Tommasi, R. Oral, Health effects and toxicity mechanisms of rare earth elements –– Knowledge gaps and research prospects, Ecotoxicol. Environ. Saf. 115 (2015) 40–48, https://doi.org/10.1016/j.ecoenv.2015.01.030

[37] S. Hirano, K.T. Suzuki, Exposure, metabolism, and toxicity of rare earths and related compounds, Environ. Health Perspect. 104 (1996) 85–95, https://doi.org/10.1289/ehp.96104s185

[38] T. Shigematsu, Lanthanum carbonate effectively controls serum phosphate without affecting serum calcium levels in patients undergoing hemodialysis, Ther. Apheresis Dial. 12 (1) (2008) 55–61, https://doi.org/10.1111/j.1744-9987.2007.00541.x

[39] E. Willbold, X. Gu, D. Albert, K. Kalla, K. Bobe, M. Brauneis, C. Janning, J. Nellesen, W. Czayka, W. Tillmann, Y. Zheng, F. Witte, 3 Effect of the addition of low rare earth elements (lanthanum, neodymium, cerium) on the biodegradation and biocompatibility of magnesium, Acta Biomater. 11 (2015) 554–562, https://doi.org/10.1016/j.actbio.2014.09.041

[40] A.N. Campbell, W.H.W. Wood, The silver-calcium system, Can. J. Chem. 49 (8) (1971) 1315–1316, https://doi.org/10.1139/v71-216

[41] L. Lutterotti, Total pattern fitting for the combined size-strain-stress-texture determination in thin film diffraction, Nucl. Instrum. Methods Phys. Res. Sect. B: Beam Interact. Mater. Atoms 268 (2010) 334–340, https://doi.org/10.1016/j.nimb.2009.09.053

[42] R.J. Gillies, N. Didier, M. Denton, Determination of cell number in monolayer cultures, Anal. Biochem. 159 (1) (1986) 109–113, https://doi.org/10.1016/0003-2697(86)90314-3

[43] W.A. Alexander, L.D. Calvert, J.A.H. Desaulniers, H.S. Dunsmore, D.F. Sargent, The silver-calcium phase diagram, Can. J. Chem. 47 (1969) 611–614, https://doi.org/10.1139/v69-093

[44] M.R. Baren, The Ag-Ca (Silver-Calcium System), Bull. Alloy Phase Diagr. 9 (1988) 228–231, https://doi.org/10.1007/BF02881269

[45] K.A. Gschneidner, F.W. Calderwood, The Ag-La (Silver-Lanthanum System), Bull. Alloy Phase Diagr. 4 (1983) 370–374, https://doi.org/10.1007/BF02868082

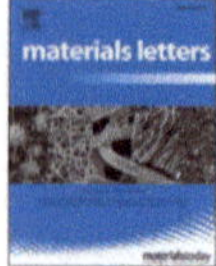

Contents lists available at ScienceDirect

Materials Letters

journal homepage: www.elsevier.com/locate/matlet

Bioresorbable AgCe and AgCeLa alloys for adapted Fe-based implants

Jan Tobias Krüger [*], Kay-Peter Hoyer, Mirko Schaper

Chair of Materials Science, Paderborn University, Mersinweg 7, 33100 Paderborn, Germany

ARTICLE INFO

Keywords:
Silver alloys
Corrosion
Biomedical application
Bioresorbable metal
Iron-based alloys

ABSTRACT

Implants often overtake body function just for a certain time and remain as an unnecessary foreign body or have to be removed. Thus, resorbable implants are highly beneficial to reduce patient burden. Besides established materials, Iron-(Fe)-based alloys are in focus due to superior mechanical properties and good biocompatibility. However, their degradation rate needs to be increased. Phases with high electrochemical potential could promote the dissolution of residual material based on the galvanic coupling. Silver (Ag) is promising due to its high electrochemical potential (+0.8 V vs. SHE), immiscibility with Fe, biocompatibility, and anti-bacterial properties. But to prevent adverse consequences the Ag-particles, remaining after dissolution of the matrix, need to dissolve. Thus, a bioresorbable Ag-alloy is required. Regarding the electrochemical potential and degradation behavior of binary alloys, Cerium (Ce) and Lanthanum (La) are well-suited considering their biocompatibility and antibacterial behavior. Accordingly, this research addresses AgCe and AgCeLa alloys as additives for Fe-based materials with adapted degradation behavior. Furthermore, degradable Ag-alloys combined with inert implant materials could enable the controlled release of antibacterial active Ag-ions.

1. Introduction

Since Fe is an essential trace element for the human body it is biocompatible. Furthermore, Fe-based alloys show superior mechanical properties and moderate degradation rate compared to established resorbable materials like Magnesium (Mg)-based alloys. Consequently, Fe-based alloys are in focus as bioresorbable implants. However, a modification is necessary due to the insufficient degradation rate [1,2,3,6].

Alloying, e.g. with Mn, leads to a reduction of the electrochemical potential and promotes the evolution of phases with different potentials, resulting in enhanced degradation. But due to the formation of passivating surface-layers, the degradation rate still needs to be increased [1,3]. Based on the formation of galvanic cells and the resulting anodic dissolution of the material with a less noble potential, the formation of local areas with noble character enhances the dissolution [3–5]. Such phases can be formed by Fe-insoluble elements with noble potential. Precious metals, like Ag, added to iron-based biodegradable materials act as effective cathodes leading to increased degradation [3–6].

The immiscibility of Fe and Ag prevents mixing of the components and consequently enables the evolution of phases with high electrochemical potential but also makes the processing challenging. Thus, new powder-based manufacturing methods like laser beam melting (LBM) have to be applied, which enables the processing of FeMn for resorbable implants and 316L for antibacterial active implants with Ag-alloys due to small melt pool and strong melt-pool movements [3,5,7,8].

Ag is well suited as it is used in medicine for a long time and can prevent implant infection. But a large intake of Ag is discussed critically and the amount of silver intake is a key factor for biocompatibility. Since the World Health Organization proposes a lifetime intake of up to 10 g Ag as uncritical and an osteosynthesis plate with a weight of 2.0 g containing 5.0 wt-% silver releases only 100 mg Ag, which is uncritical and the advantages prevail [7,9,10]. Nevertheless, the Ag-phases released during the degradation may lead to adverse consequences as thrombosis. Therefore, the silver phases need to consist of a bioresorbable Ag-alloy.

Various Ag-alloys are examined [11,12]. Unpublished work indicates promising results for alloys with rare earth metals like Ce and La showing uniform degradation with adapted degradation rate. Rare-earth elements are not essential and are discussed controversially, too. But the introduced amount is much less than the critical values. Furthermore, rare earth metals are successfully investigated as a component of degradable Mg-alloys, and an antibacterial effect is discussed [13,14].

To promote the degradation of Ag-alloys through a galvanic coupling, finely structured phases are promising. Therefore, a hypereutectic alloy (AgCe) with a content of 15 wt-% Ce is selected, due to the

* Corresponding author.
E-mail address: krueger@lwk.upb.de (J.T. Krüger).

https://doi.org/10.1016/j.matlet.2021.130890
Received 24 February 2021; Received in revised form 11 August 2021; Accepted 12 September 2021
Available online 15 September 2021

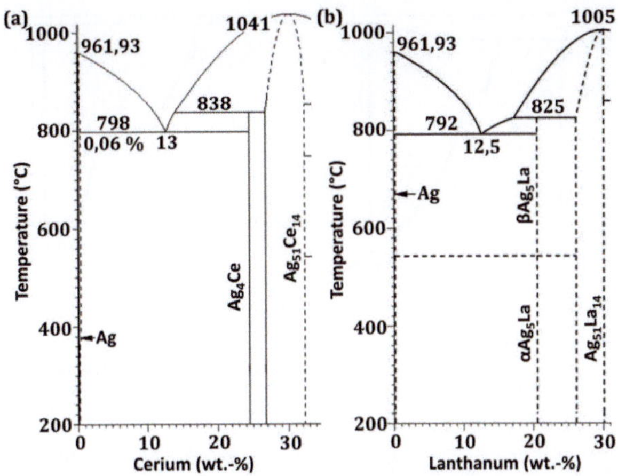

Fig. 1. Relevant sections of the phase diagrams of the (a) Ag-Ce- and (b) Ag-La-systems, modified according to [15]

Table 1
Average composition of the alloys measured by EDS.

Alloy – Microstructure component	Composition (wt.%)		
	Ag	Ce	La
AgCe - Average	84.4	15.6	—
AgCe - Island-like	78.8	21.2	—
AgCe - Eutectic	85.8	14.2	—
AgCeLa - Average	80.8	14.8	4.4
AgCeLa - Lamellae	75.1	19.6	5.3
AgCeLa - Eutectic	86	10.6	3.4

fine structure of eutectic and to prevent hypoeutectic poorly degradable Ag-phases (Fig. 1). Since, alloying with La results in loose degradation products, La can promote the uniform degradation of Ce connected to the formation of adhesive degradation products. Therefore, ternary alloy (AgCeLa) with 15 wt-% Ce and 5 wt-% La is selected. Probably the ternary system forms phases comparable to the binary system, as Ce and La can be replaced by each other since they have similar properties and are completely soluble [15].

2. Experimental procedures

The experiments were carried out as described in [16].

3. Results and discussion

3.1. Microstructural characterization

As expected, the binary AgCe consists mainly of finely structured eutectic with some prior solidified intermetallic Ce-rich island-like phases, confirmed by chemical composition and cooling curve (Table 1, Fig. 2). A hypereutectic Ce-content is measured and the cooling is slightly delayed over 800 °C linked to the formation of intermetallic hypereutectic phases. The stopping-point at 797 °C is connected to eutectic solidification. Due to fast cooling and non-equilibrium conditions, the stopping-point is not ideally shaped. Regarding the EDS-analysis the limited accuracy due to the influence of neighboring phases must be considered. Nevertheless, the measured value of 21.2 wt-% Ce for the island-like structure corresponds to the intermetallic.

AgCeLa consists of lamellae structure with high Ce- and La-contents which gaps are filled with a eutectic-like structure. Due to the similarity of binary phase diagrams the lamellae and one phase of the eutectic are probably intermetallic phase with Ce substituted partly by La, confirmed by a summarized La and Ce content of 24.9 wt-% matching the content of intermetallic in AgCe- and AgLa-systems (Fig. 1). Accordingly, the cooling is delayed between 870 °C and 795 °C connected to the solidification of hypereutectic lamellae followed by the typical stopping-point of the eutectic reaction. Higher contents of intermetallic phases are expectable due to higher combined Ce- and La-content. Thus, the amount and shape of the intermetallic phases are the main difference between both alloys.

3.2. Degradation behavior

Regarding the OCP both, Ag-alloys are suitable to adapt the degradation of FeMn. Both have higher OCPs than FeMn and could increase the degradation of Fe-based matrix material due to galvanic coupling, but only AgCeLa can be combined with 316L as AgCe has a higher OCP than 316L (Table 2). The degradation develops similarly regarding mass loss until 700 h. After 700 h the weight of AgCe is nearly constant due to the formation of a corrosion inhibiting layer (Fig. 3(a)). Consequently, the degradation process of AgCe stops, and the complete dissolution of AgCe is questionable.

In contrast, the mass of AgCeLa increases after 700 h due to the formation of corrosion products connected to a partly crack-formation indicating further degradation. The cracks develop along with the lamellae structure due to varying electrochemical potential of phases causing enhanced degradation in the contact zone between the intermetallic lamellae and the eutectic structure. The observed sample decays after immersion (Fig. 3(b)). Thus, the desired disruption of corrosion products is achieved and complete degradation of AgCeLa alloy is possible. Unfortunately, the decay could cause the release of particles in undesired sizes. For adaption, a reduced combined Ce- and La-contend is promising to reduce the amount of intermetallic phase and prevent the lamellar microstructure.

4. Conclusions

Due to biocompatibility and antibacterial behavior, AgCe and

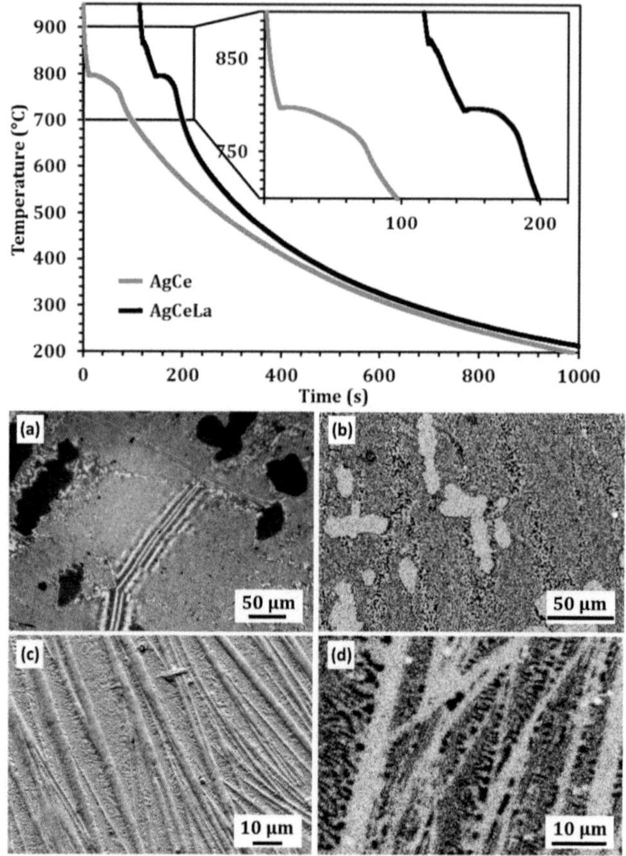

Fig. 2. Cooling curves and microstructure: (a) light microscopy of AgCe, (b) EDS-Mapping of AgCe (bright areas indexing high Ce-content), (c) SEM-SE image of AgCeLa and (d) EDS-Mapping of AgCeLa (bright areas indexing high Ce- and La-content).

Table 2
Open circuit potential in Ringer's Lactate solution vs. SHE.

Alloy	OCP (mV)
AgCe	70
AgCeLa	−170
FeMn22C0.6	−500
316L	50

AgCeLa alloys are selected as potential additives for Fe–based alloys with adapted degradation. The OCP of both alloys is promising, but the degradation of AgCe is insufficient due to a layer-formation. As expected, the addition of La breaks this inhibiting layer and enables further degradation, which unfortunately is connected to crack formation. Thus, AgCeLa is probably bioresorbable but further efforts are necessary to adapt the microstructure and prevent decay along intermetallic lamellae. As LBM influences the microstructure, an investigation of the presented or slightly adapted AgCeLa-alloy as an additive for Fe-based alloys is promising. Due to the immiscibility of Fe and Ag, the Ag-alloy creates phases inside the iron-based matrix resulting in galvanic coupling and dissolution of the matrix-material. As the Ag-phases with higher electrochemical potential remain they should degrade after the matrix comparably to the results of this study.

CRediT authorship contribution statement

Jan Tobias Krüger: Writing – original draft, Writing – review & editing, Methodology, Investigation. **Kay-Peter Hoyer:** Writing – review & editing, Project administration. **Mirko Schaper:** Project administration, Funding acquisition.

Declaration of Competing Interest

The authors declare that they have no known competing financial interests or personal relationships that could have appeared to influence

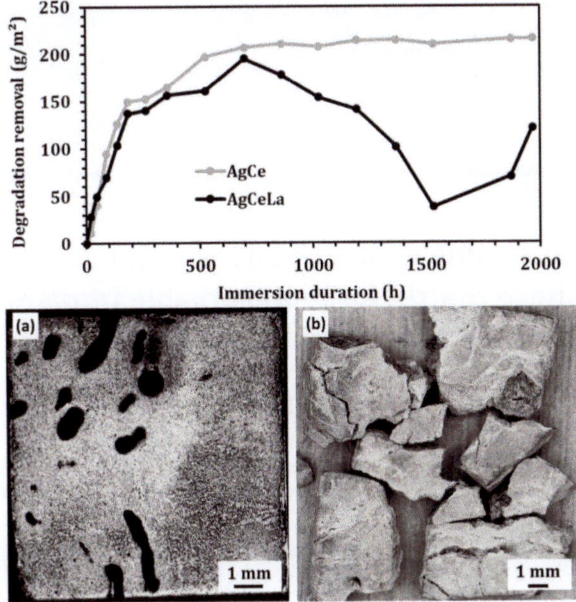

Fig. 3. Degradation rates of AgCa- and AgCaLa-alloys in Ringer's Lactate solution for 2000 h and light microscopy of degraded samples of (a) AgCe (pores due to casting imperfections) and (b) AgCeLa.

the work reported in this paper.

Acknowledgment

The authors gratefully thank the German Research Foundation (DFG) for financial support (SCHA1484/44-1).

References

[1] M. Peuster, C. Hesse, T. Schloo, C. Fink, P. Beerbaum, C. von Schnakenburg, Long-term biocompatibility of a corrodible peripheral iron stent in the porcine descending aorta, Biomaterials 27 (28) (2006) 4955–4962, https://doi.org/10.1016/j.biomaterials.2006.05.029.

[2] N. Abbaspour, R. Hurrell, R. Kelishadi, Review on iron and its importance for human health, J. Res. Med. Sci. 19 (2) (2014) 164–174.

[3] T. Niendorf, F. Brenne, P. Hoyer, D. Schwarze, M. Schaper, R. Grothe, M. Wiesener, G. Grundmeier, H.J. Maier, Processing of new materials by additive manufacturing: iron-based alloys containing silver for biomedical applications, Metall. Mater. Trans. A 46 (7) (2015) 2829–2833, https://doi.org/10.1007/s11661-015-2932-2.

[4] T. Huang, J. Cheng, D. Bian, Y. Zheng, Fe-Au and Fe-Ag composites as candidates for biodegradable stent materials, J. Biomed. Mater. Res. B Appl. Biomater. 104 (2) (2016) 225–240, https://doi.org/10.1002/jbm.b.33389.

[5] M. Wiesener, K. Peters, A. Taube, A. Keller, K.-P. Hoyer, T. Niendorf, G. Grundmeier, Corrosion properties of bioresorbable FeMn-Ag alloys prepared by selective laser melting, Mater. Corros. 68 (10) (2017) 1028–1036, https://doi.org/10.1002/maco.201709478.

[6] P.S. Bagha, M. Khakbiz, S. Sheibani, H. Hermawan, Design and characterization of nano and bimodal structured biodegradable Fe-Mn-Ag alloy with accelerated corrosion rate, J. Alloy. Compd. 767 (2018) 955–965, https://doi.org/10.1016/j.jallcom.2018.07.206.

[7] E. Zhang, X. Zhao, J. Hu, R. Wang, S. Fu, G. Qin, Antibacterial metals and alloys for potential biomedical implants, Bioact. Mater. 6 (8) (2021) 2569–2612, https://doi.org/10.1016/j.bioactmat.2021.01.030.

[8] J. Quan, K. Lin, D. Gu, Selective laser melting of silver submicron powder modified 316L stainless steel: Influence of silver addition on microstructures and performances, Powder Technol. 364 (2020) 478–483, https://doi.org/10.1016/j.powtec.2020.01.082.

[9] World Health Organization, Silver in Drinking-water Background document for development of WHO Guidelines for Drinking-water Quality Guidelines for drinking-water quality, 2nd ed. Vol. 2. Health criteria and other supporting information. WHO, Geneva, 1996.

[10] K. Mijnendonckx, N. Leys, J. Mahillon, S. Silver, R. Van Houdt, Antimicrobial silver: uses, toxicity and potential for resistance, Biometals 26 (4) (2013) 609–621, https://doi.org/10.1007/s10534-013-9645-z.

[11] K.-P. Hoyer, M. Schaper, Alloy design for biomedical applications in additive manufacturing, in: TMS 2019 148th Annual Meeting & Exhibition Supplemental Proceedings, Springer International Publishing, 2019, pp. 475–484, https://doi.org/10.1007/978-3-030-05861-6_44.

[12] A. Andreiev, K.-P. Hoyer, O. Grydin, Y. Frolov, M. Schaper, Degradable silver-based alloys, Materialwissenschaft & Werkstofftechnik. 51 (4) (2020) 517–530, https://doi.org/10.1002/mawe.201900191.

[13] G. Pagano, M. Guida, F. Tommasi, R. Oral, Health effects and toxicity mechanisms of rare earth elements - Knowledge gaps and research prospects, Ecotoxicol. Environ. Saf. 115 (2015) 40–48, https://doi.org/10.1016/j.ecoenv.2015.01.030.

[14] D. Liu, D. Yang, X. Li, S. Hu, Mechanical properties, corrosion resistance and biocompatibilities of degradable Mg-RE alloys: a review, J. Mater. Res. Technol. 8 (1) (2019) 1538–1549, https://doi.org/10.1016/j.jmrt.2018.08.003.

[15] ASM Alloy Phase Diagram Database. (1999).

[16] J.T. Krüger, K.-P. Hoyer, V. Filor, S. Pramanik, M. Kietzmann, J. Meißner, M. Schaper, Novel AgCa and AgCaLa alloys for Fe-based bioresorbable implants with adapted degradation, J. Alloy. Compd. 871 (2021), 159544, https://doi.org/10.1016/j.jallcom.2021.159544.

Available online at www.sciencedirect.com

jmr&t
Journal of Materials Research and Technology

journal homepage: www.elsevier.com/locate/jmrt

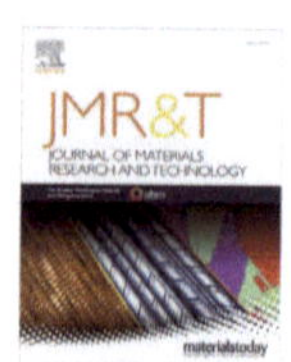

Original Article

Formation of insoluble silver-phases in an iron-manganese matrix for bioresorbable implants using varying laser beam melting strategies

Jan Tobias Krüger [a,b,*], Kay-Peter Hoyer [a,b], Florian Hengsbach [a,b], Mirko Schaper [a,b]

[a] Chair of Materials Science, Paderborn University, Mersinweg 7, 33100 Paderborn, Germany
[b] DMRC – Direct Manufacturing Research Center, Paderborn University, Mersinweg 3, 33100 Paderborn, Germany

ARTICLE INFO

Article history:
Received 31 March 2022
Accepted 1 June 2022
Available online 7 June 2022

Keywords:
Laser beam melting
Bioresorbable metal
Iron-based alloys
Insoluble alloys
Melt flow

ABSTRACT

The combination of immiscible metals is targeted to fulfil the requirements for certain technical and biomedical applications. The degradation behavior of iron (Fe) is adjustable by the insertion of silver (Ag) phases with high electrochemical potential causing anodic dissolution of the Fe-based matrix. To process such immiscible materials, powder-based processes like laser beam melting (LBM) can be applied. The continuous fusion process characterized by a small melt pool, strong melt flow, and rapid solidification enables the creation of well-dispersed Ag phases. To enable the adaption of insoluble phases this study aims to understand the mechanism responsible for the incorporation of insoluble phases within the matrix. Based on the characterization of FeMn with 5 wt.-% Ag processed via LBM the surface tension is determined as the essential influencing factor and a model of the melt pool including melt flow is developed. The Ag forms an almost closed layer on the top surface with an increased thickness between adjacent melt tracks. The lower surface tension of Ag compared to FeMn keeps Ag on the top and the outward-directed melt flow of underlying FeMn transports the Ag to the sides of the melt pools. During the generation of the next slice, deep parts of the Ag layer remain in the bulk material. By adaption of process conditions, e. g. the hatch distance, the size and shape of Ag phases are influenced. Moreover, the chemical composition depends on the LBM strategy, since the diffusion of alloying elements is differently pronounced.

1. Introduction

The progress in medical technology enables the treatment of increasingly more diseases and injuries by implants [1–3].

Accordingly adapted materials for implants are needed to fulfil the specific requirements of each application. Thus, the actual research focuses on the development of new materials utilizing the potential of the materials. For example, an adapted stiffness reduces stress shielding effects, antibacterial properties prevent

* Corresponding author.
   E-mail address: krueger@lwk.upb.de (J.T. Krüger).
https://doi.org/10.1016/j.jmrt.2022.06.006

implant infections, and bioresorbable materials make operations for implant removal redundant or prevent adverse consequences caused by implants remaining in the human body [1,4–10].

The material discussed in this paper mainly addresses the bioresorbability but also antibacterial properties. A few degradable implants based on magnesium (Mg) or polymers are already in successful clinical use [5,6,11–13]. But these materials do not fit all possible applications of bioresorbable implants e.g. due to low mechanical load capability [2,12,14,15]. Consequently, there is a need for the development of further degradable materials. Iron is a promising material due to its good biocompatibility, suited mechanical properties, and low degradation rate compared e.g. to Mg-based materials [5,12,16–19]. Nonetheless, pure Fe is not suitable and an adaption is necessary to overcome limitations like the too low degradation rate and reduced mechanical properties compared to commercially used Fe-based alloys [4,5,20,21] A promising approach is the adaption of the material properties via alloying, whereby manganese (Mn) is auspicious [16–18,22]. Mn reduces the electrochemical potential of iron and can additionally adjust local differences in the electrochemical potential due to Mn segregations. Both effects increase the degradation rate of FeMn alloys compared to pure Fe, as materials with less noble character are typically increasingly attacked by corrosion, and differences in the electrochemical potential can enhance the anodic dissolution of less noble areas [21,23–27]. The mechanical properties are improved by alloying with Mn as well. In combination with small amounts of carbon (C), the strength and ductility can be improved due to twinning induced plasticity (TWIP) [15,21,28–31]. Additionally, promising results have already been obtained with regard to the biocompatibility of FeMn alloys [18,23,32–35].

Nevertheless, the degradation rate of FeMn alloys is still too low for most applications [23,27,31,33,36]. Hence, an approach is the targeted creation of phases with high electrochemical potential within the FeMn to increase the degradation rate due to the anodic dissolution of FeMn which is less noble [30,36–38]. Noble elements are well suited as they have high electrochemical potential whereby especially silver is promising due to its insolubility with iron, well-investigated biocompatibility, and relatively low price, compared to platinum or palladium [39–42]. Additionally Ag addresses the increasing occurrence of multiresistant germs and resulting in heavily treatable infections due to its antibacterial properties [3,43,44]. Various studies show enhanced degradation rates by the addition of Ag [17,30,39,45,46]. In consideration of the results from Huang et al. [39], Niendorf et al. [30], and Liu et al. [47], an amount of 5 wt.-% Ag is selected for the present study as it is a good compromise between enhanced degradation, mechanical properties, and biocompatibility. Since pure Ag-phases would remain after degradation of the iron-based matrix, a modification of Ag-phases is necessary. Degradable Ag-alloys are promising as these alloys enable the degradation of Ag-phases while their electrochemical potential is still high enough to promote anodic dissolution of the matrix material [42,48–50].

The immiscibility of Fe and Ag on the one hand enables the existence of Ag-phases within a FeMn matrix but, on the other hand, makes the processing of such material challenging [51–53]. A possibility to overcome this limitation is to use powder metallurgical processes, as the compounds can be mixed mechanically [39,54,55]. Aside from sintering, LBM is a promising technique to gain homogenously distributed Ag-phases within a Fe-matrix, due to the interaction of the small melt pool, strong melt movements, and rapid melting and solidification [30,56–58]. For the successful application of LBM, the formation of Ag-phases must be controllable as their morphology and chemical composition is crucial for the biocompatibility and degradation behavior.

In particular, processing degradable Ag alloys in combination with a Fe matrix LBM is beneficial, as melting and solidification occur extremely fast. Thus, only a short time for diffusion of alloying elements between both melts exists. However, the diffusion of alloying elements between insoluble liquid metals can be very distinctive [59,60]. The enrichment of Mn in the Ag-phase would be problematic as it could reduce the open circuit potential of Ag-phases and the release of alloying elements from a degradable Ag alloy is critical as this influences the degradability. Consequently, an adaption of the LBM process to reduce the interaction respectively diffusion between the different components is addressed. Therefore, a reduced energy input should be beneficial to reduce the time in which high temperatures prevail resulting in less diffusion. The reduction of overall energy input, as well as the reduction of the energy input for every single melt track to create smaller melt pools, is to be targeted.

To control the morphology of the Ag-phases is also crucial since nanoparticular Ag is discussed critically in literature and should be avoided [58,61]. Otherwise, small and well-dispersed Ag-phases are necessary to enable homogenous degradation and only slightly affect the mechanical properties. Whereby for some applications a reduced stiffness can be beneficial for an adaption of stiffness to the elasticity of the human bone [5,62]. As Ag and Fe are immiscible, no carrying connection between the Ag-phases and the matrix is expected and the Ag-phases should act like pores regarding the mechanical properties. Hence, these phases can contribute to an adapted stiffness. Additionally, a targeted porosity can be applied to further adapt the degradation behavior and to improve the osteoconductivity [18,63].

In conclusion, the possibility for the targeted adaption of Ag-phases within a mixed material is crucial for the successful application of LBM to manufacture FeMnAg materials. Thus, this work aims to understand the mechanisms of the formation of Ag-phases within the melt pool during LBM. Therefore, mechanically mixed powders of Ag and FeMn are processed via LBM under varying process conditions to observe the Ag-phase formation under different conditions and to identify correlations between the LBM process and the phase formation. The process strategies employed address the concerns mentioned before. As the starting point, a layer thickness of 50 µm resulting in samples with a high density is selected. For comparison, a strategy with a layer thickness of 25 µm creating samples with a high density is applied to enable smaller melt pools which should lead to smaller Ag-phases and less uptake of Mn. Conclusively, a strategy with a layer thickness of 50 µm and a reduced energy input is chosen to enable a targeted porosity for further adaption of the degradation behavior and to reduce the Mn uptake by the Ag-phases.

Understanding the influence of physical effects during LBM is addressed in current research to gain knowledge regarding the continuous formation of bulk material which enables the targeted adaption of the LBM process [64–67]. However, the bulk formation characteristics vary considerably depending on the material processed, the powder properties, and the LBM process settings. Therefore, a general statement regarding the entire LBM process flow is not possible [67,68]. Furthermore, the physical effects and processes cannot be observed directly, thus they have to be investigated by melt pool observation, transfer of known physical effects, and investigation of the achieved materials' microstructure and properties. Therefore, models are generated and the LBM process is simulated with satisfying results confirming the hypothesis of the melt pool as well as the relationship between process and properties [64–66,69,70]. Following this strategy, the knowledge gained from literature is correlated with the observed microstructure and mechanical properties of the additively manufactured FeMnAg specimens to determine the mechanisms that are decisive for the formation of the Ag-phases. Afterwards, a qualitative model is developed to demonstrate the possibilities to influence the Ag-phases via the LBM process. Aside from the LBM process itself, the feedstock properties like chemical composition, the proportion of both powder materials, and the morphology of the powders are decisive. But, since the final manufacturing process is the LBM process, the feedstock has to be adapted to the LBM process. E.g. the vaporization or the diffusion of alloying elements can be compensated by increasing the amount of these elements in the starting material [71,72]. Thus, the knowledge of the LBM process is even crucial for the adaption of the feedstock.

The successive formation of bulk material during LBM caused by the laser is dominated by an interplay of the sequentially proceeding sub-processes respectively physical effects. The main processes that occur continuously as the laser is moving along the melt tracks are absorption of laser energy, melting, consolidation of the melt, melt fluxes within the melt pool, heat dissipation, and solidification [64,66,67,69]. These sub-processes interact with each other and induce further effects like spattering, whereas the sub-processes themselves depend on further physical effects [67,69,73,74] e.g., the adsorption of laser radiation depends on the absorptivity of the materials in the molten and solidified state, on the reflection within the powder or within an eventually formed keyhole. Additionally, the deflection by the vapor flow influences the introduced amount of energy, which is decisive for the laser radiation of the powder or the melt [75–78]. Thus apparently a large number of material properties, process conditions, and interacting physical effects have to be considered to explain the melt pool and the resulting structure of the final material.

Now to develop a model, the main influencing processes have to be identified and incorporated, subsequently further influencing variables can be added to refine the model. There is consensus in the literature on the major influencing physical effects although different shapes of melt pools and melt fluxes are reported [64–66,69,76]. Both, melt fluxes and the resulting shape of the melt pool, are mainly influenced by the surface tension resulting in Marangoni fluxes and evaporation of material in the laser spot resulting in recoil pressure [64,70,75,79,80]. Here, the intensity of evaporation, in particular, determines the welding mode and the shape of the melt pool. At a low recoil pressure, heat transfer is determined by conduction and is referred to as conduction mode. As the recoil pressure increases, a depression of the melt is formed resulting in a keyhole welding mode, in which the heat is transferred predominantly by convection. Between both modes, there is a continuous transition with different proportions of the modes, which is called transition mode. The mode can be identified by the shape of the melt pool [64,76,81,82]. Depending on the material and process parameters all modes can occur for LBM, although for iron-based materials a transition mode is most common [74,77,81,82]. Other, less influencing factors are gravity, buoyancy force, drag force, thermal conductivity, radiation, and any phase transformation or chemical reaction that might occur [64,66,70,79].

Evaporation and the resulting welding mode are crucial for the formation of the Ag-phases but another effect is expected to be more effective. As the formation of Ag-phases occurs in the molten state the fluxes within the melt pool and the solidification are significantly influencing the phase evolution. Following Khairallah et al. [64], the melt pool can be divided into sections with different prevailing forces. The first part of the melt pool before and in the area of the laser spot is mainly determined by the recoil pressure in the keyhole or transition mode. The following transition and tail end region are mainly dominated by surface tension affecting the melt flow and surface of the already solidified melt tracks [64,66,80]. As the shape of Ag-phases is defined during solidification and their position results from previous melt flow, the transition and tail-end regions are prevailing. Thus the evolution of the Ag-phases is expected to be mainly influenced by the surface tension respectively the resulting Marangoni effect. Accordingly, the course and differences of surface tension are discussed in detail and a model based on effects caused by surface tension is developed. This model explains the observed morphology of the Ag-phases and enables the further targeted development of the LBM process.

The intended application for the immiscible FeMnAg alloys is discussed critically [33,34,83–85]. However, the process understanding generated within the presented study can be transferred to other immiscible material combinations and applications like bearings with soft lead (Pb) phases [51]. Combining degradable Ag-alloys with a non-degradable Fe-based matrix to achieve a defined release of antibacterial acting Ag ions from inert implants to prevent implant-related infections is a further possible application [49,55,86,87].

2. Experimental procedures

The feedstock used for LBM is argon-gas-atomized FeMn powder with a nominal composition of 22 wt.-% Mn, 0.6 wt.-% C and Fe bal. (Nanoval GmbH & Co. KG, Berlin, Germany) and pure Ag powder (ECKART TLS, Bitterfeld-Wolfen, Germany).

The nominal particle size is 20 μm–63 μm for both powders. Via a Mastersizer 2000 (Malvern Panalytical GmbH, Kassel, Germany) the particle size distribution (PSD) is measured. The adsorption of the laser radiation (wavelength of 1070 nm) by the powder deposited like in the powder bed is determined with an HL-2000-HP-FHSA (Ocean Optics, Inc., Orlando, Florida, USA).

For the production of the samples, pure FeMn and a mixture of FeMn with 5 wt.-% Ag powder is employed. In the following, the manufactured specimens are referred to as FeMn and FeMnAg. To achieve a homogenous mixture of both powders a drum hoop mixer with half-filled powder containers is utilized for 30 min at 60 rpm. Before processing via LBM, the powder is vacuum dried to guarantee a humidity of less than 5%. The samples are produced with an SLM 280^{HL} (SLM Solutions Group AG, Lübeck, Germany) in an argon atmosphere (4.6) with an oxygen level of less than 0.03%. The conducted parameters (Table 1) are selected based on previous, unpublished work and are used for FeMn as well as FeMnAg identically. All samples (cubes: 10 × 10 × 10 mm³, cylinders for computer tomographic (CT) analysis: ø 5 × 10 mm³, blocks for tensile specimens: 13 × 36 × 10 mm³) of each material composition are produced with one built-job. Thus, no differences between samples should occur due to different process conditions. The process strategies are named according to the slice thickness and achieved density (High Density: HD and Low Density: LD).

To increase the isotropy, for all strategies, a rotation of 67° between the scanning direction of the following layers is performed. In addition, to gain an impression of the melt pool shape and the evolving silver phases, a few samples are produced without rotation. These samples are marked with "-0" following the enumeration, e.g. 50-HD-0.

For tensile tests, miniature dogbone specimens with a measuring length of 8 mm, a width of 3 mm, a thickness of 1.5 mm, and a radius of 2 mm are used. The design of the specimen follows DIN EN ISO 6892-6 but does not fulfil the criterium of a proportional specimen. From bulk material produced in a horizontal alignment with 50-HD parameter setting the specimens are eroded and subsequently ground with silicon carbide (SiC) paper with the final step P2500.

For microstructural investigations, the samples are ground iteratively up to a final step with P4000 whereby a least 1 mm material is removed to prevent the influence of edge effects. Subsequently, vibration polishing with a mixture of 50% silica polishing suspension with a grit size of 50 nm (Cloeren Technology GmbH, Wegberg, Germany) and 50% of corrosion inhibiting lubricant (CUTLUB (Diluted, 3%), Cloeren Technology GmbH, Wegberg, Germany) is carried out for 12 h. Some of the samples are etched with 0.5% HNO$_3$-acid for 30 s up to 150 s.

For the light microscopic (LM) investigations, a Keyence VHX5000 digital microscope (KEYENCE Corporation, Osaka, Japan) and an Axiophot (Carl Zeiss AG, Oberkochen, Germany) are used. Further investigations are carried out with a field emission scanning electron microscopy (FE-SEM) Zeiss Ultra Plus (Carl Zeiss AG, Oberkochen, Germany) with secondary electron (SE) and backscatter electron (BSE) detectors. Additionally, the SEM is equipped with an electron backscattered diffraction (EBSD) detector DIGIVIEW 5 (AMETEK, Berwyn, Pennsylvania, USA) and energy-dispersive X-ray spectroscopy (EDS) detector Octane Pro (AMETEK, Berwyn, Pennsylvania, USA). All presented EBSD-data are not cleaned. The three-dimensional structure of the Ag-phases is observed with a micro-CT SkyScan 1275 (Bruker Corporation, Billerica, Massachusetts, USA). Via optical emission spectroscopy (OES) with Q4 TASMAN (Bruker Corporation, Billerica, Massachusetts, USA) the chemical composition of the LBM samples is determined.

Quasistatic tensile tests were performed with an MTS 858 Table Top System (MTS Systems Corporation, Eden Prairie, Minnesota, USA) equipped with an extensometer using a testing velocity of 0.13 mm/min which corresponds to a strain rate less than 0.0025 1/s following DIN EN ISO 6892-6.

3. Results

The investigations start with the characterization of powder properties and distribution of Ag-particles within the powder bed, as they are crucial for LBM [88–91]. According to (Fig. 1a,b,e) both powders have a suitable (PSD) and morphology, thus a sufficient flowability can be assumed. The Ag-powder shows more distinctive satellites while the FeMn-powder has a slightly smaller average PSD. Although minor differences in the deposition of particles with different sizes can occur along the coater path, a homogenous distribution of both materials can be assumed over the hole powder bed, as the difference in particle size is neglectable [88]. To verify the homogeneity of the Ag distribution, powder samples are collected from the powder bed after processing. The powder is fixed by instant glue (Fig. 1c). Investigation via SEM proves a suitable stochastical mixture of both powders (Fig. 1d). Thus, no segregation of Ag occurs during powder handling or recoating. Such a powder bed with a homogenous mixture of FeMn and Ag absorbs just 61.8% of the laser energy in contrast to a powder bed containing only FeMn absorbing 69.3%. A separation of Ag is also during LBM not expectable as the investigation of the oversize powder sieved out after the LBM shows no enrichment of Ag (Fig. 1g,h). Noticeable is an arrangement of Ag around the FeMn-particles and the bonding of FeMn particles by Ag. This observation is explainable due to the different melting points. Consequently, it can be assumed that spattering does not cause excessive ejection of Ag and no reason for a deviating Ag-content in the final part is obvious.

Within the FeMn powder internal pores are detected (Fig. 1f) which can result in a higher porosity in the final parts fabricated via LBM [91–93].

FeMn and FeMnAg exhibit similar properties in terms of density and shape of the melt pools for identical parameter

Table 1 – Parameters for processing via LBM.			
Strategy	50-HD	25-HD	50-LD
Slicing in μm	50	25	50
Scanning velocity in mm/s	750	785	850
Laser power in W	280	195	240
Hatching in μm	110	115	190

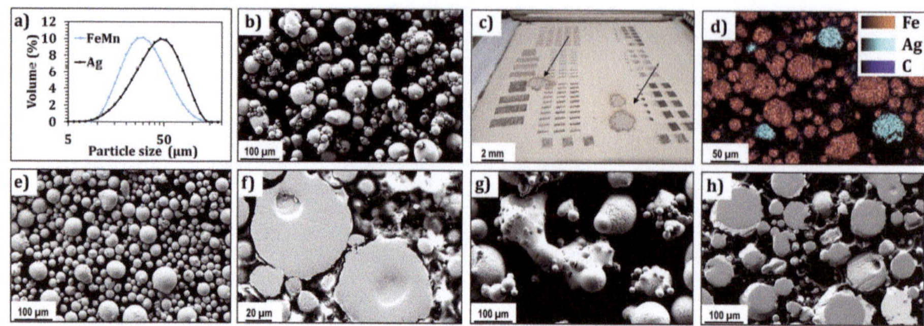

Fig. 1 — a) PSD of FeMn- and Ag-powders; b) SEM-SE: Ag-powder; c) Sampling from the powder bed; d) SEM-EDS: Distribution of Ag-Particles in the powder bed; e,f) SEM-SE: FeMn-powder; g,h) SEM-SE: Oversize particles (>100 μm) sieved out after the LBM process.

settings. Furthermore, an appropriate distribution of Ag-phases is obtained without any adaption of parameters. Thus, both materials can be processed with the same parameters even though Ag influences the process conditions due to differences in melting point, specific enthalpy, reflectivity, etc.

Examination of the top surfaces in the as-built condition (Fig. 2a,c) enables the observation of a surface comparable to surfaces that are covered with powder while recoating and then re-melted during the generation of the next layer. Almost the entire top surface consists of Ag interrupted only by some FeMn and some predominately Mn and O rich areas. A certain Mn content is found even in the predominantly Ag-containing areas. Most of the FeMn detected on the top surface is unmolten FeMn-powder that adheres to the Ag. The heat of the melt causes a rise of gas which leads to a lateral gas influx that transports FeMn particles from the lateral powder bed into the melt [73,74,76,94]. These particles, therefore, do not result from melt pool motion and are not relevant for the determination of the melt fluxes.

Areas with high Mn and O content are distributed over the entire surface and enriched along the centerline of the melt tracks (Fig. 2a). Since only just small amounts of other elements are found in these areas and as the phases are at the top surface, these areas are slag consisting mainly of manganese oxides (MnO) [95–98]. Hence, the powder of the following layers is placed on an almost continuous silver layer with some slag on the top.

According to Fig. 2b,d both components are a uniform part of the side surface. However, an amount of 9 wt.-% Ag is detected at the side which is almost twice the value for the homogenous powder mixture.

The previous results are confirmed by the continuous Ag-layers on the top surface of samples built with the 25-HD and 50-HD strategy (Fig. 3a-h). In comparison, the Ag-layer is most pronounced in the 50-HD strategy, and the thickness of the Ag-layer tends to be increased at the overlap points of adjacent melt pools (Fig. 3a-c). In addition, (Fig. 3a,c) reveals the remaining Ag-phases in the bulk material at the overlap

points of the melt tracks. As Ag forms a continuous layer on the top surface, Ag agglomerates and the Ag-phases in the bulk are formed from this Ag layer and not from individual powder particles.

Samples with high porosity have just a few Ag-phases on the top surface (Fig. 3i,j), white arrows). Thus, in a process strategy leading to a porous structure, some Ag is directly incorporated into the bulk material.

Also, the shape of melt pools is similar for FeMn and FeMnAg, a difference in the depth of the melt pools is present. The 50-HD-0 and 25-HD-0 strategies result in a depth of 110 μm–150 μm and 40 μm–60 μm for FeMnAg which is slightly less than for FeMn exhibiting a melt pool depth of 120 μm–150 μm resp. 60 μm–80 μm. As the measured melt pool depth is significantly more than the slice thickness, a re-melting of the Ag-layer occurs. However, certain areas of the Ag-layer at the overlapping areas of the melt pools are likely to be deep enough below the top surface and could not be re-melted.

To characterize the morphology of the Ag-phases, Fig. 4 gives an overview of samples built with different LBM strategies. Before considering the Ag-phases, a close look at the samples with targeted high density showing small round pores is provided. These pores are suspected to be filled with gas which can originate from different sources. The inclusion of gas in the feedstock powder as shown for FeMn before (Fig. 1f) might be a reason [91–93]. But as demonstrated by Niendorf et al. [30], a pore-free fabrication of FeMnAg via LBM is possible. Thus, these pores are not caused by the addition of Ag and, due to the small size and acceptable amount, no effect regarding the formation of Ag-phases is expected.

The morphology of the Ag-phases varies depending on the process strategy, but even within samples prepared using the same strategy, the shape of Ag-phases varies. For example, (Fig. 4a,b,e,f) highlights the different morphology of Ag-phases from samples built with the 50-HD strategy. Two different types of the shape of Ag-phases can be identified. One of these types are phases with an elongated shape perpendicular to the building direction (Fig. 4e). As the strategy foresees a rotation

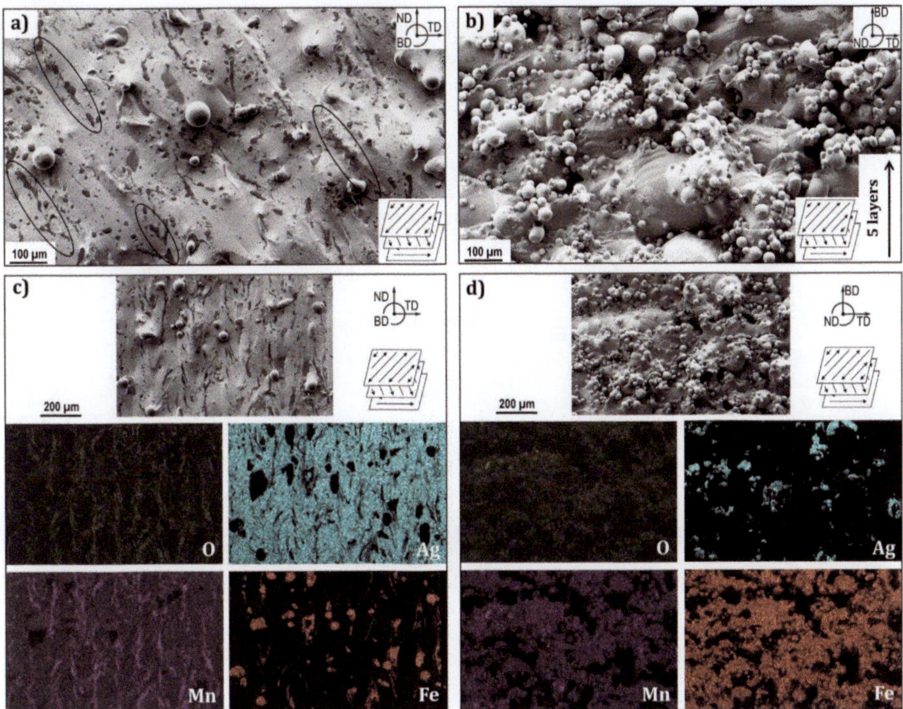

Fig. 2 – As-built surfaces of FeMnAg built with 50-HD strategy: a) SEM-SE: Melt tracks on the top surface, deposits at the center of the melt track; b) SEM-SE: Side surface with adhering powder particles; c) SEM-SE with corresponding SEM-EDX: Enrichment of Ag on the top surface; d) SEM-SE with corresponding SEM-EDX: Side surface.

of 67° between the following layers elongation of Ag-phases even in the depth of the sample is expected. Thus, the Ag-phases are spread flat parallel to the slices of the LBM process. However, the Ag is not uniformly distributed, as Ag is enriched in some slices perpendicular to the building direction. Since the distance between the enriched planes is much larger than the height of the single layers, this enrichment of Ag is not related to the single layers of the LBM process.

The Ag-phases of the second type are located at the overlap of two melt pools. They can be regularly arranged, and in some cases, the Ag is drawn along the melt pool boundary towards the lower sharp edges (Fig. 4b,f,j). The regular arrangement marked in Fig. 4g also reveals the arrangement between adjacent melt pools. These Ag-phases have a more compact shape, which has a favorable effect on mechanical properties. Whereas the arrangement of Ag between the melt pools is more pronounced for the 50-LD-0 strategy, as Ag is located almost between all melt tracks (Fig. 4h,l). Considering the absence of a continuous Ag-layer on the top of the porous samples (Fig. 3i,j), the strategy 50-LD facilitates the incorporation of Ag into the bulk material

and favors the arrangement along the sides of the melt pools. If a rotation of the scanning direction is applied, the distribution of Ag and pores seems to be more randomly (Fig. 4d) since the regular arrangement of tracks is not present anymore.

Other randomly distributed Ag phases that do not correspond obviously to the previously mentioned structures are also observed. The influence of the LBM strategy is most evident for the 50-LD strategy, but samples built with 25-HD and 50-HD also differ. Ag phases of 25 HD (Fig. 4c,k) have a more homogeneous shape, are more evenly distributed and appear to be slightly smaller on average. Here, the shape of the Ag phases in sections perpendicular to the building direction (Fig. 4i,k) tends to be rounder and more compact than in sections parallel to the building direction. A comparable shape of Ag phases is observed by Quan et al. [57] and Niendorf et al. [30] as well.

The three-dimensional examination by CT, including a larger data set, confirms the conclusions previously drawn from the cross-sections. Homogenous distribution of Ag-phases is observed (Fig. 5). Whereby the distribution can be

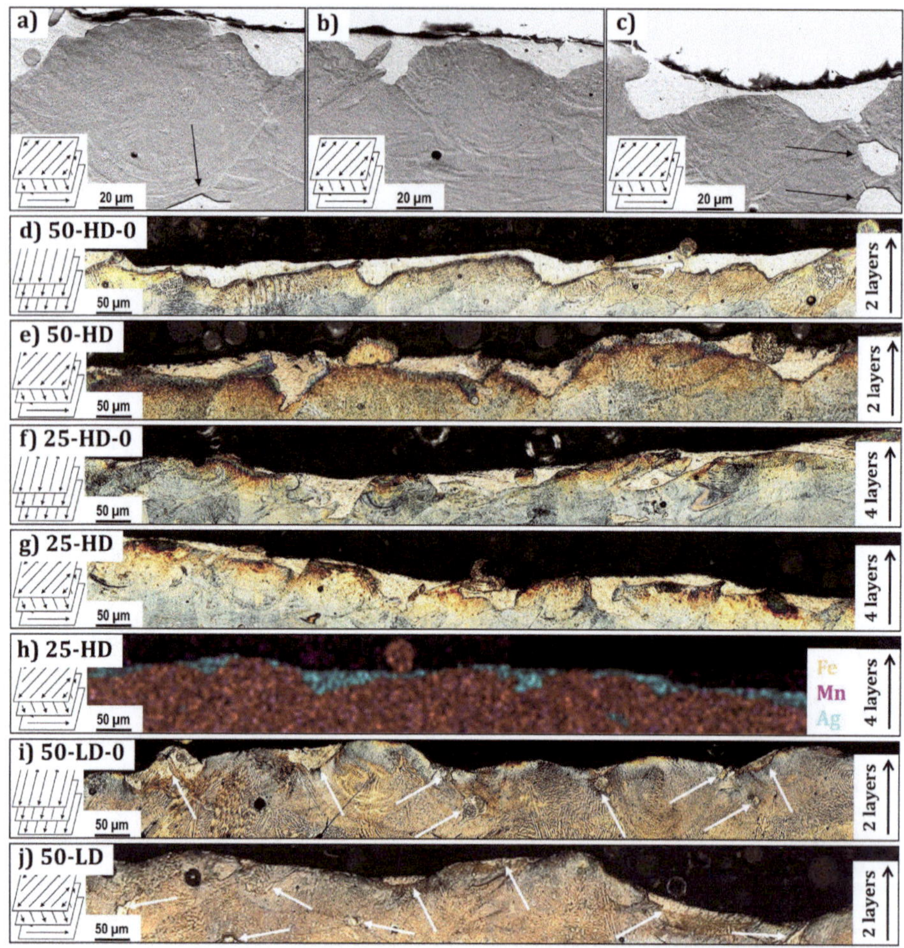

Fig. 3 – Enrichment of Ag at the top surface of FeMnAg, etched cross-sections: a,b,c) SEM-SE: 25-HD, incorporated Ag-phases marked with arrows; d-g) LM, almost closed Ag-layer on the top; h) SEM-EDX; i,j) LM, Ag-phases indicated with arrows.

depicted better in the attached videos. The Ag-phases of the 25-HD strategy (Fig. 5c) are more homogeneously distributed than those of the 50-HD strategy (Fig. 5d) as in the 50-HD sample are less but bigger Ag-phases. This qualitative effect is confirmed by the quantitative size distribution of the Ag-phases (Fig. 5a). However, each material contains Ag-phases larger than the Ag particles of the feedstock, confirming the coagulation of Ag during the LBM. Thus, the formation of Ag-phases from the Ag-layer on the top surface is conclusive. Furthermore, the data verify that Ag is more likely incorporated into the porous structure without the formation of

extensive Ag-layers on the top surface, as the smaller Ag-phases indicate less coalescence compared to the other strategies.

The two types of Ag phases identified previously are also confirmed, as a layered structure perpendicular to the building direction and Ag phases lined up along the scan direction are present (Fig. 5e,f). Analysis of the CT data reveals an average Ag content of 4.8 vol.-% Ag-phases. For determination of the Ag-content, the additional volume of the Ag-phases due to the inclusion of an average of 10.6 wt.-% Mn, discussed in the following section, must be considered. Thus, a content of

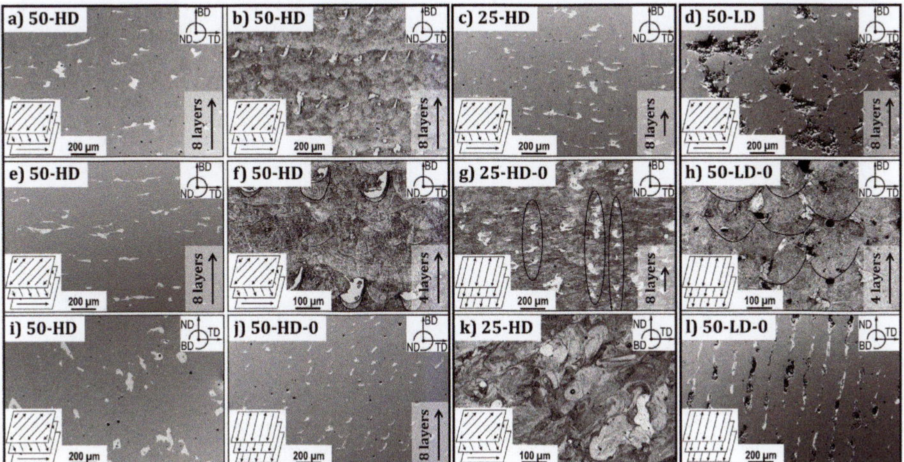

Fig. 4 – Distribution of Ag-phases within the FeMnAg bulk material: a,c-e,i,j,l) SEM-SE; b,f-h,k) LM of etched samples; b,f) Regular distributed Ag-phases located at the overlapping point of melt pools; e) Layers enriched with Ag perpendicular to the building direction; g) In building direction superimposed Ag-phases; h) Ag-phases between adjacent melt pools; l) Ag-phases orientated along the scanning direction.

5.6 wt.-% Ag is determined via CT and confirms that the Ag-content within the bulk material corresponds with the content in the powder feedstock. The minor difference might be due to measurement deviations e.g. caused by the selection of the threshold value for the identification of Ag.

The microstructure and the chemical composition of the matrix material and the Ag-phases are investigated, as they influence the degradation behavior and allow conclusions on the solidification process. The FeMn Matrix solidifies in a columnar-dendritic manner (Fig. 6a), which is typically

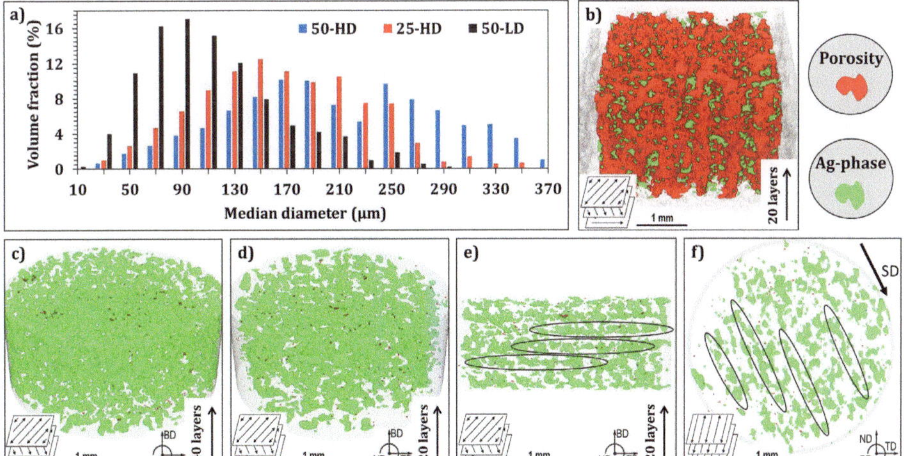

Fig. 5 – 3D-distribution of Ag-Phases analyzed via CT: a) Size distribution of Ag-phases; b) 50-LD; c) 25-HD; d) 50-HD; e) 50-HD: Layer-wise distribution of Ag-phases; f) 50-HD-0: Ag-phases distributed along the scanning direction (SD).

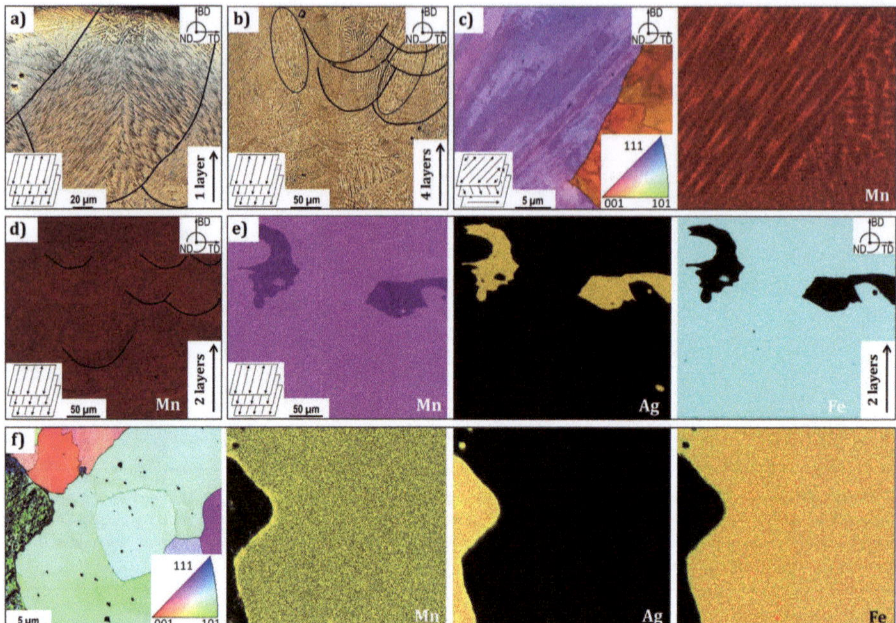

Fig. 6 — Solidification structure and Mn-segregations: a) LM: FeMn 50-HD-0, columnar-dendritic solidification; b) LM: FeMn 25-HD-0, regular shape of melt pools, epitaxial columnar structure; c) SEM-EBSD with corresponding SEM-EDS: FeMn 50-HD, segregation along with columnar-dendritic structure; d) SEM-EDS: FeMn 50-HD-0, segregation of Mn along the melt pool boundaries; e) SEM-EDS: FeMnAg 50-HD f) SEM-EBSD with corresponding SEM-EDS, diffusion heat-treated FeMnAg with Mn-enrichment along the FeMn-Ag-phase boundary.

associated with the segregation of alloying elements [72,99]. Combined EBSD- and EDX-data acquisition confirms the segregation of Mn along with the solidification structure (Fig. 6c). In addition, the comparison of the EBSD- and EDX-data reveal the expected mutual influence of grain structure and segregations [65,77,100,101]. The columnar dendritic structure propagates epitaxially across the melt pool boundaries (Fig. 6b).

But segregation of Mn also occurs on a larger scale as the structure of the melt pools is irregular concerning the Mn distribution (Fig. 6d). Consequently, the alloying elements are segregated regularly along with the melt pools [102–104]. These segregations may contribute to an increase in the rate of degradation due to the locally different electrochemical potential [23,25,64]. Measurements via OES indicate a chemical composition of 0.58 wt.-% C, 20.31 wt.-% Mn and Fe bal. for FeMn processed by the 50-HD strategy. Thus, the content of Mn is reduced compared to the starting material probably due to combustion during the LBM process, but the Mn- and C-content are still sufficient to achieve a completely austenitic structure [31,56,105,106].

As indicated by EDX (Fig. 1c), the former pure Ag contains Mn (Fig. 6e). The Mn-content of the Ag-phases differs depending on the LBM strategy applied. On average, the Ag-phases of samples built with the 50-HD strategy contain 12.7 wt.-%, whereas the strategies 25-HD and 50-LD result in only 11.5 wt.-% respectively 7.5 wt.-% Mn. While the Mn content of the Ag phases from high-density samples differs only slightly, the Mn content of the phases in porous material varies between 2 wt.-% and 12 wt.-%. Due to the lack of Mn in the Ag feedstock, the Mn must be transferred from the FeMn to the Ag by the interaction of both materials during LBM. In this process, the Mn content in the Ag phases exceeds the equilibrium content at room temperature of 6 wt.-% [52]. Thus, the uptake of Ag occurs in the molten state, and the rapid solidification prevents concentration equilibrium so that the Ag is supersaturated with Mn.

As mentioned in the introduction, the uptake of Mn can be critical, so a diffusion heat treatment slightly below the melting point of the Ag-phases (940 °C/24 h, cooling: 20 K/h, vacuum) is performed, which allows a concentration equilibration [52]. For samples built with the 50-HD strategy, the Mn content of the Ag is reduced by this heat treatment to 6.5 wt.-% which corresponds to the equilibrium at room temperature. Otherwise, the heat treatment results in a balance of Mn segregations and a slight enrichment of Mn along the phase

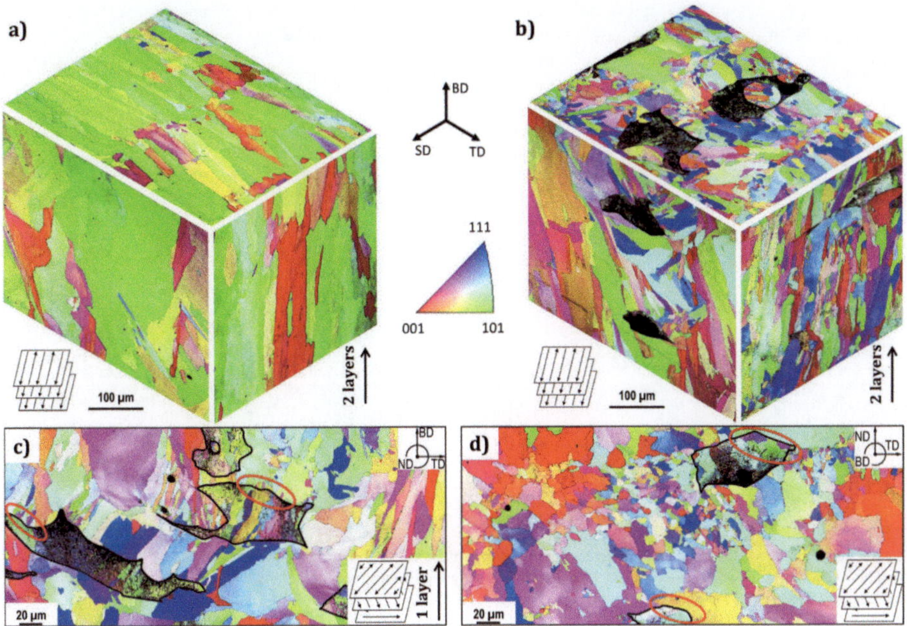

Fig. 7 — SEM-EBSD: Inverse pole figures, crystallographic direction referred to the building direction: a) FeMn 50-HD-0, regular structure with preferred <101> orientation; b) FeMn+5Ag 50-HD-0, irregular structure, Ag-phases dark-colored due to reduced image quality c,d) FeMnAg 50-HD, no obvious correlation between the orientation of Ag and FeMn at the phase boundaries, Ag-phases bordered.

boundary between matrix and Ag-phases (Fig. 6f). As the austenitic structure of the matrix remains, the heat treatment does not change the matrix significantly.

Measurements of the open circuit potential (OCP) performed as described by Krüger et al. [42] confirm a reduced potential by the uptake of Mn. For pure Ag, a potential of +170 mV versus SHE is measured in Ringer's Lactate Solution. While for an alloy containing 6 wt.-% and 20 wt.-% Mn a potential of +150 mV vs. SHE respectively +58 mV vs. SHE is determined.

The grain structure allows conclusions with respect to the solidification process. Therefore, FeMn and FeMnAg built with the 50-HD-0 strategy are investigated (Fig. 7a,b). The structure of FeMn is regularly and epitaxial with preferred <101> orientations along the building direction, as it is typical for this parameter setting [77,101,104,107]. This epitaxial structure results from the alternating growth during the solidification of adjacent and following melt tracks. The energetically preferred growth direction depends not only on the temperature gradient, as the epitaxial growth in a direction varying from the temperature gradient can be energetically favorable compared to the formation of new grains [77,101,108]. Thus, homogenous solidification conditions result in a regular grain structure. As this regular structure and preferred orientation are not observed for FeMnAg, the solidification conditions for FeMnAg differ and are less regular. As expected, (Fig. 7c,d) indicate the absence of any correlation between the orientation of the FeMn and the Ag, since grains of the Ag-phases are adjacent to grains of the FeMn-phase with different orientations.

To enhance the microstructural investigation, the results of tensile tests are presented in Fig. 8. The results are in line with the expectations, as the mechanical properties of FeMnAg are reduced compared to FeMn. In particular, the elongation at break is significantly lower, as the minimum elongation for FeMnAg is 16% compared to 25% for FeMn. The addition of Ag reduces the tensile strength by 17.5%. The only slightly reduced yield strength and stiffness are comprehensible as the Ag-phases reduce the load-bearing cross-section by only a few percentages. Additionally, the ductile material behavior contributes to the suitable load capacity of FeMnAg, since stress peaks can be relieved by the flow of the material and the notch effect is reduced [30].

The ductile fracture behavior is confirmed by the fracture surfaces, as the typical honeycomb fracture structure is observed (Fig. 8c,d). The ductile character is more pronounced

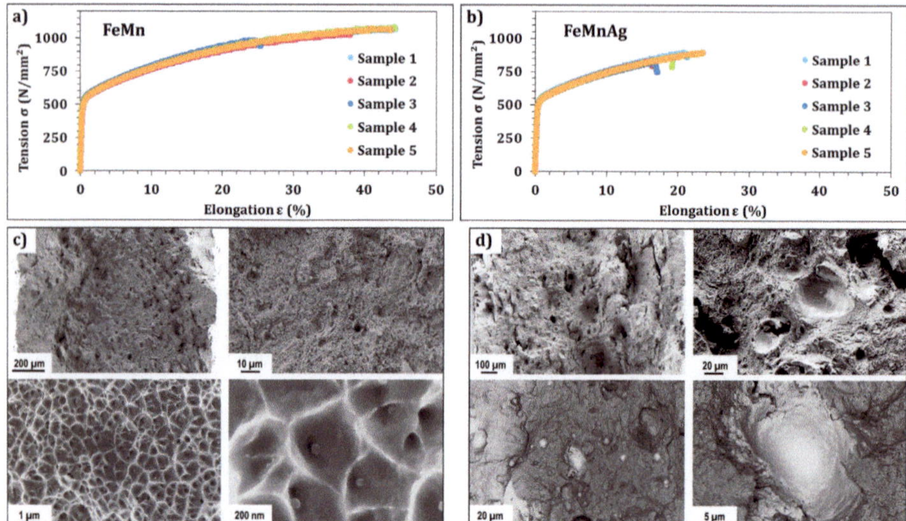

Fig. 8 – Results of tensile tests: a,c) Tension elongation diagram for FeMn and corresponding SEM-SE of fraction area, ductile fraction behavior; b,d) Tension elongation diagram for FeMnAg and corresponding SEM-SE of fraction area, reduced loading capacity, detachment of Ag-phases.

in FeMn, since the fracture surface of FeMnAg is more cleaved. These cleavages arise by the transition of the fracture line from one Ag phase to another, as they are the weakest points in the material.

Apart from the mechanical properties, the analysis of the fracture surface of FeMnAg confirms that the Ag-phases have no connection to the matrix since the Ag-phases are detached from the matrix. Fig. 8d) shows both detached Ag-phases as well as depressions where the Ag is detached. Thus, the mechanical properties, as well as the fracture surface, confirm that Ag-phases do not contribute to mechanical load capacity.

4.   Discussion

The discussion will provide an understanding of the interaction of FeMn with Ag during LBM and identify the critical parameters that influence the formation and thus, the resulting morphology of the Ag-phases. Subsequently, the suitability of FeMnAg processed via LBM as a bioresorbable material is evaluated.

First, the welding mode is identified to determine the main influencing physical effects and the character of the melt flow. King et al. [82] defined the keyhole welding mode for melt pools with a depth larger than half the width of the melt pool. Following this definition e.g. Fig. 6a) indicates the keyhole mode. However, as stated in the introduction, the definition of an additional transition mode is purposeful and by comparing the shape of the melt pools with the results presented in the

literature, the transition mode can be identified for FeMn and FeMnAg [74,76,77,81]. Consequently, the shape of the melt pool and the melt flow are determined by depression due to the recoil pressure and Marangoni flow due to the surface tension [64,70,75,79,80].

The recoil pressure for FeMn is mainly influenced by Mn due to the high vapor pressure of Mn. Evaporation of Mn is confirmed by the reduced Mn content of the samples manufactured with LBM [56]. Although the boiling temperature and enthalpy of evaporation of Ag and Mn differ only slightly, only a small contribution of Ag to the recoil pressure is expected as the vapor pressure is lower for Ag than for Mn [105,109–112]. The similar shape of the melt pools of FeMn and FeMnAg confirms that no significant difference in evaporation and consequently in welding mode occurs. Thus, the influence of vaporization is not decisive for the formation of the Ag-phase, particularly as the formation occurs at end of the melt pool where the Marangoni fluxes are dominant (cf. Introduction).

The continuous formation of bulk material starts with the absorption of the laser energy. As the size of the melt pools does not differ significantly, a similar energy absorption can be assumed. However, the slightly shallower depth of the melt pools of FeMnAg is conclusive due to the high reflectivity and thermal conductivity of Ag. In addition to the adsorption behaviour, the thermal conductivity, viscosity, etc. influence the melt flow and the shape of the melt pool [113–115]. The energy is absorbed by the powder, non-consolidated melt, and mainly by the melt within the melt pool [73,75,76,82]. Absorption by powder and non-consolidated melt with the

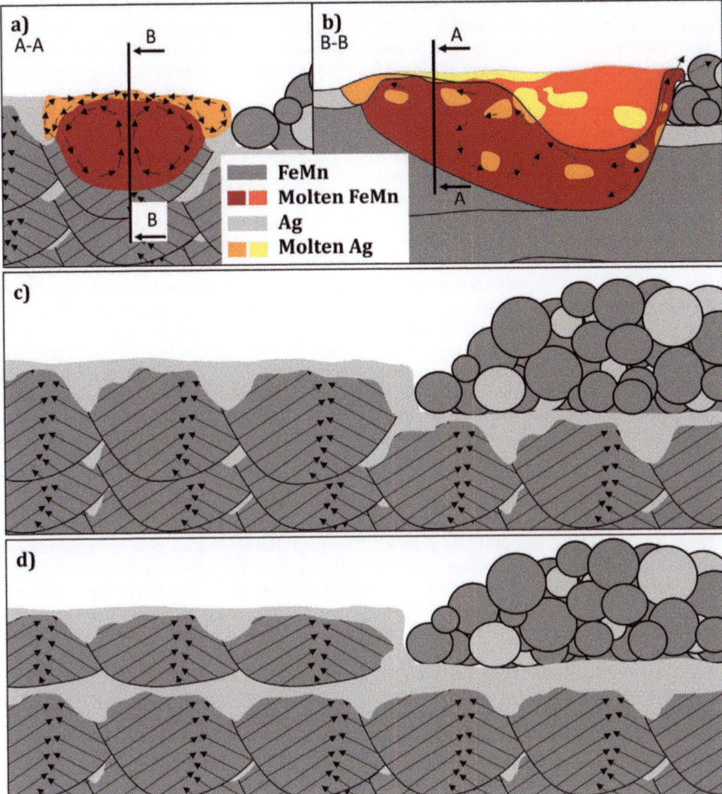

Fig. 9 — Schematic model of the formation of Ag-phases during the LBM process for a strategy without rotation between the scanning directions of subsequent layers; a,b) Fluxes of Ag within the FeMn, separation of Ag at the top of the melt pool, the shape of the melt pool and melt fluxes according to [64–66,69,76]; c) Formation of Ag-phases at the overlap point of adjacent melt tracks; d) Formation of flat elongated Ag-phases perpendicular to the building direction.

unchanged distribution of Ag is less affected, as adsorption of FeMnAg is only 7.5% lower than for FeMn [75]. However, the Ag-layer on top indicates a drastic reduction of energy adsorption, which does not occur in this way. Thus, the Ag-layer is not present in the zone of interaction with the laser. This is conclusive, as the interaction between laser and melt occurs in the depression zone, which is determined by the vapor pressure being similar for both phases [64,75,76]. Thus, the molten Ag emerges into the melt pool in the depression zone and the coagulation at the top occurs in the melt pool behind this region. This intense emerging of Ag into the FeMn melt is even conclusive considering the pronounced uptake of Mn by the Ag.

The similar shape of the melt pools of FeMn and FeMnAg indicates that the addition of Ag does not significantly change the melt flow. Hence, a schematically model of the melt pool and melt flow can be obtained from the literature addressing single-phase melts (Fig. 9a). Depending on the welding mode and the melt pool shape, the melt flows are adapted from the literature [64–66,69,76]. Different flows are reported in the literature, e.g. regarding vortex formation and direction of melt flow [64,75,116,117]. Nonetheless, the relevant vortexes can be determined by evaluating the present results.

As previously suggested, the coagulation of Ag at the surface occurs in the part of the melt pool where the surface tension is decisive. The coagulation due to surface tension is easily explainable, while the arrangement of Ag on the top surface requires in-depth discussion. The coagulation of Ag is conclusive as the surface tension supports processes that reduce interfaces, and FeMn and Ag exhibit a distinctive phase boundary as indicated by the optical investigation, REM-EDX, REM-EBSD, and tensile tests. Due to the immiscibility, the absence of any bonding is to be expected. But alloying elements in insoluble melts may cause some type of bonding [118]. However, the absence of increased Mn content at the phase boundaries (Fig. 6e) confirms the absence of bonding [118].

At the surface of liquids, the surface tension causes a force from areas with low surface tension to areas with high tension [97,119]. Accordingly, it is essential to determine the course of surface tension and the difference in surface tension between the two phases to understand the melt flow. Surface tension is a temperature-dependent material property, strongly influenced by the chemical composition. Thus, impurities, segregations, and the adsorption of small amounts of oxygen (O) affect the surface tension [97,119–121]. Consequently, the surface tension ratio between the two phases can only be estimated. Fe has a fundamentally higher surface tension than Ag, and alloying with Mn and C reduces the surface tension [119,122–124]. However, the surface tension is still likely to be higher than for pure Ag, in addition, a reduction of surface tension by uptake of Mn is conclusive for Ag. However, the adsorption of O influences the surface tension of both melts strongly, with the influence being more pronounced at low O partial pressure for Fe [119,122,123,125]. Although the LBM is performed under argon-atmosphere, the remaining O content is high enough to cause changes in surface tension [119,125,126]. However, the adsorption of O is a time-dependent process, and the time in the molten state is significantly short during LBM [119,121]. Additionally, the FeMn melt is covered by Ag, thus low adsorption of O by FeMn is conclusive and higher surface tension of FeMn can be assumed.

As surface tension attracts liquids with low surface tension to areas with high surface tension, the accumulation of Ag-phases in contact with the top surface on the FeMn is conclusive. From the assumption of higher surface tension for FeMn, it follows that the surface tension between Ag and FeMn as well as between Ag and the surrounding atmosphere is lower than between FeMn and the surrounding atmosphere. Thus, additional surface energy would be required to move FeMn in a position above the Ag. However, gravity counteracts the arrangement of Ag at the top, but as the surface-driven forces are strong due to very steep temperature gradients during LBM, there is no significant influence of gravity [64,66,70,79]. Consequently, the Ag remains at the top of the melt. The argumentation above can be also applied the other way around: The position of Ag at the top indicates lower surface tension of Ag. Therefore, the assumptions regarding surface tension are consistent.

The effects discussed previously explain why Ag, once it reaches the surface, remains attached to the top, but not how the Ag is transported to the top. First, the Ag-phases can reach the surface randomly as they move around with the melt flow since the drag force keeps the Ag-phases in the melt flow even against gravity [98,127,128]. But a further effect, caused by surface tension, supports the upward movement of Ag. As the temperature at the bottom is lower than at the top center of the melt pool, the surface tension is higher under floating Ag-phases than above them. Consequently, the surrounding FeMn flows from the top of the Ag-phase under the Ag-phase and pushes the Ag upward [129,130].

Aside from the surface tension, the different melting points of Ag and FeMn contribute to the Ag-layer on the top. As the FeMn solidifies before Ag, the Ag-melt is pushed ahead of the solidification front like segregations of alloying elements [64,104,131]. But this effect is not decisive since the Ag is not positioned in the finally solidifying upper center of the melt track.

The gradient of the surface tension within the melt is crucial for identifying the direction of the Marangoni flow. Both, the Ag and the Fe show a typical decrease in surface tension with increasing temperature [119,120,125]. Consequently, the common melt flow from the center to the sides of the melt pool is conclusive due to the highest temperature in the center of the melt pool [88,97,117,120]. But the adsorption of oxygen can change the direction of the melt flow, as the adsorption is more pronounced for lower temperatures and thus, decreases the surface tension at the sides of the melt pool [95,97,120]. However, for FeMn, a minor influence of the adsorption is likely, and an outward vortex for the FeMn melt can be assumed (Fig. 9a). Furthermore, the outward flow is conclusive as it explains the transport of Ag to the sides of the melt pool and the emerging of Ag on the sides of the melt pools. The deposition of Ag on the sides of the melt pool is even confirmed by the enrichment of Ag at as-built side surfaces.

Regarding the Ag, an increased influence of O adsorption can be expected due to the more intense contact with the atmosphere and a change in direction of the vortex is possible. Additionally, the Ag is pulled to the side of melt pools by the underlying FeMn, and due to the need for closed circular flow, an inward flow in the upper part of the molten Ag-layer can be assumed. The inward flow of Ag is confirmed by the position of the slag in the middle of the melt tracks [95,98]. The slag is positioned at the top due to low surface tension and low density [96–98]. Thus, the directions of melt flow shown in Fig. 9(a and b) are expected.

Following the consideration of the melt fluxes and the behavior of Ag within the melt pool, the formation of Ag-phases from the Ag-layer is explored. It has to be taken into account that even under stable process conditions, the continuous formation of bulk material varies significantly as e.g. the depth of the melt pool can vary up to 20% [104]. Additionally, the results reveal more non-uniform conditions due to the addition of Ag. Thus, regularities of the formation of Ag-phases are disturbed by varying conditions, and Ag is incorporated by random processes. Nevertheless, two types of Ag-phases can be identified (cf. results).

A development of defects by enhancement of irregularities over a few slices is known for LBM [64,70,79,80,89]. According to this mechanism, coagulation of Ag results in an increasing Ag-layer on the top of FeMn over a few slices. Apparently, the incorporation of Ag into the bulk becomes easier with increasing layer thickness. When a threshold is exceeded, the layer is too thick for complete re-melting and some Ag remains. Since the accumulation of Ag at the sides of the melt pools increases the layer thickness at the overlap points, the remaining of Ag-Phases is more likely at these overlap points. Fig. 9c) demonstrates this effect, resulting in the formation of Ag-phases of the second type (cf. results) for a process strategy without rotation between subsequent layers. If a rotation is conducted, the deep center of the melt pool of the subsequent layer completely re-melt the existing Ag-layer even at the overlap points and the continuous tracks of Ag along the scanning direction are interrupted. Thus, the desired Ag-phases with a compact shape can be obtained.

The Ag-phases with a flat spreading shape are not re-melted parts of Ag-layers. However, this shape cannot be explained by the accumulation of Ag alone, as the formation of the second type, or random Ag-phases, must be assumed before an Ag-layer is thick enough to remain on large scale. Consequently, the depth of the melt pools for certain layers must be reduced enabling the formation of such phases (Fig. 9d). As these Ag-phases only exist within certain layers, the random process variations cannot be used as an explanation. However, the interaction between scan direction and gas flow is a suitable explanation. Scanning parallel to the gas flow may reduce the depth of melt pools since a vapor plume transported parallel to the laser spot causes increased adsorption and ablation of laser power [77,78]. This hypothesis is affirmed by the absence of flat spread Ag-phases in samples without rotation of scan direction. Consequently, adjusting the process strategy to disable the scanning along the gas flow can prevent the formation of these large Ag-phases. The prevention of these Ag-phases is necessary as they have a large effective diameter like lack of fusion defects [91].

Aside from the morphology of Ag-phases, the intensity of the interaction between the melts resulting in uptake of Mn has to be addressed. Reducing the interaction that leads to changes in the intended chemical composition is necessary, as it limits the targeted adaption of the chemical composition of the Ag-phases. However, the uptake of Mn by Ag should be acceptable for the intended application as the electrochemical potential of the Ag-phases is still higher than for FeMn with a potential of −500 mV [42] and Mn might support the dissolution of the Ag-phases.

However, the determined Mn contents fit the observed morphology of Ag-phases, since the most pronounced interaction can be assumed for the 50-HD strategy due to the most distinctive coagulation of Ag on the top. In accordance, the highest Mn-content is observed for the 50-HD strategy, indicating also the most distinctive interaction. The Ag-phases obtained with the 25-HD strategy are characterized by smaller and more homogenous Ag-phases, indicating simplified incorporation of Ag, and the Ag-phases correspondingly contain over 1 wt.-% less Mn. In the 50-LD strategy, some of the Ag is incorporated directly without coagulation at the top surface. Accordingly, some Ag-phases with very low Mn-content are detected. Thus, part of the Ag of the porous structure interacts only briefly, while another part of the Ag interacts more intensively with the Mn-melt, resulting in different Mn-contents in the melt. This is conclusive as for the 50-LD strategy no distinctive but partial layer of Ag is observed on the top surface. Consequently, some Ag is incorporated directly into the bulk material without being part of an Ag-layer on the top before.

Thus, the results presented in this study reveal that a reduction in interaction and adjustment of morphology can be addressed simultaneously. The promotion of second-type Ag phases facilitates the incorporation of Ag, which leads to a lower uptake of Mn by Ag. Apparently, the hatch distance has a significant influence, as an increasing hatch distance reduces the melting of Ag deposited at the sides of the previous adjacent tracks. However, sufficient bonding between adjacent tracks is required, and the hatch distance cannot be increased arbitrarily. Accordingly, further parameters have to be taken into account. Promising results are obtained by reducing the layer thickness in particular regarding the morphology as the Ag-phases are more homogenous and smaller. The reason for this effect is the smaller melt pool, which enables the re-melting of only thinner Ag-layers. Additionally, the hatch distance is crucial with respect to the size of the melt pool, since the size of the melt pool is reduced while the hatch distance is slightly increased compared to the 50-HD strategy.

For the 25-HD strategy, however, the uptake of Mn by the Ag is only slightly reduced, and even for the 50-LD strategy, the uptake of Mn can only be suppressed for some Ag-phases. Thus, the possibility of adjusting the morphology of the Ag-phases via LBM parameters has already been successfully demonstrated, while the uptake of Mn is still unsatisfactorily high for all strategies applied and further efforts are needed to address this challenge. It is unlikely that the interaction can be prevented by parameter adjustment alone since even the 50 LD strategy shows pronounced Mn uptake by some Ag phases, although pronounced porosity is accepted for this strategy. Whereas, the porosity might be even beneficial in adapting the mechanical and corrosion properties. One way to reduce the interaction could be to change of welding mode, as the distinctive mixing of phases occurs within the depression zone. Thus, a parameter adaption resulting in the conduction mode with less influence of the recoil pressure might result in a faster formation of the Ag-layer without distinctive mixing. But then, the Ag-layer might be problematic due to increased reflection of the laser radiation.

Therefore, further possibilities have to be considered. The uptake or loss of alloying elements can be integrated into the intended application. For example, the Mn in the Ag-phases could contribute to the dissolution of the Ag-phases. Or the changes in the chemical composition during LBM can be compensated by adjusting the feedstock. Whereas the usage of smaller Ag-particles as feedstock for achieving smaller Ag-phases in the bulk material is not effective at an amount of 5 wt.-% Ag since Ag accumulates and the Ag-phases originate from the Ag-layer. Another possibility is a subsequent heat treatment to enable concentration equilibrium. This strategy has been successfully applied for FeMnAg, but may not be effective if a degradable Ag-alloy is used, as even desired alloying elements of the Ag-alloy might diffuse into the matrix. As the use of an Ag-alloy is necessary anyway to adjust the degradation behavior, the impact on the LBM process can be considered when selecting the alloying elements. Alloying can reduce the saturation concentration for Mn in Ag and change the physical properties of the Ag-phases. For example, the reduced surface tension of the Ag-alloy could enhance the observed effects and provide increased deposition of Ag at the edges of the melt pool.

The results presented here, highlight the potential of FeMnAg alloys processed via LBM but also the remaining concerns for the use as bioresorbable material. The developed model allows an understanding of Ag-phase formation and the targeted tailoring of the LBM process. However, targeting the formation of regular Ag-phases by adapting the LBM-parameters is a concern, as the conditions fluctuate during LBM and the fluctuation is exacerbated by the addition of Ag. Consequently, an additional adjustment of the chemical composition of the Ag-phases is beneficial.

Nevertheless, FeMnAg processed by LBM is promising as bioresorbable material due to the acceptable mechanical properties (cf. Niendorf et al. [30]) and the potentially suited morphology and chemical composition of the Ag-phases. The segregations of Mn in the matrix, the reduced Mn-content in the Ag-phases of the heat-treated samples, as well as, the porosity could contribute to increased degradation. Thus, the determination of the degradation behavior is useful to further target the adaption of FeMn-AgCaLa alloy system as well as the processing by LBM.

5. Conclusions

The present study determines the mechanisms of incorporation of insoluble Ag-phases into a Fe-based matrix during LBM and identifies the influencing variables.

- The formation of Ag-phases is mainly influenced by surface tension-driven effects.
- An Ag layer is formed on the top surface.
- The outward FeMn flow transports Ag-melt to the sides of the melt pool resulting in increased Ag-layer depth at the overlap points of the melt pools.
- The Ag-phases in the bulk material originate from the Ag-layer on the top.
- The position of the Ag at the overlap points of the melt pool can be utilized to adjust the morphology of Ag-phases, e.g. by varying the hatch distance.
- During LBM, the interaction of both melts results in the diffusion of Mn into the Ag. The Mn uptake can be reduced by adapting the LBM strategy.
- Targeting the Ag-phases is challenging due to increased process variations caused by the addition of Ag.

Therefore, adapting the LBM process is challenging but the presented results enable further targeted research. The knowledge gained can be used to process FeMn in combination with a degradable Ag-alloy.

However, the results also highlight the potential of FeMnAg as a degradable material for implants. The Ag-phases are well dispersed and despite the uptake of Mn, the potential of the Ag-phases is high enough to cause anodic dissolution of the FeMn matrix. Thus, the investigation of the degradation behavior of the different materials is appropriate to determine the influence of e.g. the Ag-uptake or porosity to define requirements for the adaption of LBM.

Data availability statement

The raw/processed data required to reproduce these findings cannot be shared at this time as the data also forms part of an ongoing study.

Declaration of Competing Interest

The authors declare that they have no conflict of interest.

Acknowledgement

The authors gratefully thank the German Research Foundation (DFG) for financial support. This work results from the project SCHA1484/44-1. Furthermore, the authors also would like to thank Dr.-Ing. Anatolii Andreiev, Eva Popp, Anja Puda, Thomas Janzen and Lea Kaspersmeier for promoting the production and examination of the tested material.

REFERENCES

[1] Veerachamy S, Yarlagadda T, Manivasagam G, Kdv Yarlagadda P. Bacterial adherence and biofilm formation on medical implants: a review. Proc IME H J Eng Med 2014;228:1083–99. https://doi.org/10.1177/0954411914556137.

[2] Chen Q, Thouas GA. Metallic implant biomaterials. Mater Sci Eng R Rep 2015;87:1–57. https://doi.org/10.1016/j.mser.2014.10.001.

[3] Mauffrey C, Herbert B, Young H, Wilson ML, Hake M, Stahel PF. The role of biofilm on orthopaedic implants: the "Holy Grail"of post-traumatic infection management? Eur J Trauma Emerg Surg 2016;42:411–6. https://doi.org/10.1007/s00068-016-0694-1.

[4] Carluccio D, Demir AG, Bermingham MJ, Dargusch MS. Challenges and opportunities in the selective laser melting of biodegradable metals for load-bearing bone scaffold applications. Metall Mater Trans 2020;51:3311–34. https://doi.org/10.1007/s11661-020-05796-z.

[5] Hermawan H. Updates on the research and development of absorbable metals for biomedical applications. Progress in Biomaterials 2018;7:93–110. https://doi.org/10.1007/s40204-018-0091-4.

[6] Prakasam M, Locs J, Salma-Ancane K, Loca D, Largeteau A, Berzina-Cimdina L. Biodegradable materials and metallic implants—a review. J Funct Biomater 2017;8:44. https://doi.org/10.3390/jfb8040044.

[7] Alonzo M, Primo FA, Kumar SA, Mudloff JA, Dominguez E, Fregoso G, et al. Bone tissue engineering techniques, advances and scaffolds for treatment of bone defects. Current Opinion in Biomedical Engineering 2021;17:100248. https://doi.org/10.1016/j.cobme.2020.100248.

[8] Leordean D, Dudescu C, Marcu T, Berce P, Balc N. Customized implants with specific properties, made by selective laser melting. Rapid Prototyp J 2015;21:98–104. https://doi.org/10.1108/RPJ-11-2012-0107.

[9] Echeverrigaray FG, Echeverrigaray S, Delamare APL, Wanke CH, Figueroa CA, Baumvol IJR, et al. Antibacterial properties obtained by low-energy silver implantation in stainless steel surfaces. Surf Coating Technol 2016;3:345–51. https://doi.org/10.1016/j.surfcoat.2016.09.005.

[10] Zhang E, Zhao X, Hu J, Wang R, Fu S, Qin G. Antibacterial metals and alloys for potential biomedical implants. Bioact Mater 2021;6:2569–612. https://doi.org/10.1016/j.bioactmat.2021.01.030.

[11] Zivic F, Affatato S, Trajanovic M, Schnabelrauch M, Grujovic N, Choy KL. Biomaterials in clinical practice. Cham: Springer; 2018.

[12] Zheng YF, Gu XN, Witte F. Biodegradable metals. Mater Sci Eng R Rep 2014;77:1–34. https://doi.org/10.1016/j.mser.2014.01.001.

[13] Redlich C, Schauer A, Scheibler J, Poehle G, Barthel P, Maennel A, et al. In vitro degradation behavior and biocompatibility of bioresorbable molybdenum. Metals 2021;11:761. https://doi.org/10.3390/met11050761.

[14] Ratner BD, Hoffman AS, Schoen FJ, Lemons JE. Biomaterials science. 3rd ed. Academic Press; 2013.

[15] Hermawan H, Dubé D, Mantovani D. Development of degradable Fe-35Mn alloy for biomedical application. Adv Mater Res 2007;15–17:107–12. https://doi.org/10.4028/www.scientific.net/AMR.15-17.107.

[16] Abbaspour N, Hurrell R, Kelishadi R. Review on iron and its importance for human health. J Res Med Sci 2014;19:164–74. https://www.ncbi.nlm.nih.gov/pmc/articles/PMC3999603/.

[17] Dargusch MS, Venezuela J, Dehghan-Manshadi A, Johnston S, Yang N, Mardon K, et al. In vivo evaluation of bioabsorbable Fe-35Mn-1Ag: first reports on in vivo hydrogen gas evolution in Fe-based implants. Adv Healthcare Mater 2020;10. https://doi.org/10.1002/adhm.202000667.

[18] Mandal S, Viraj, Nandi SK, Roy M. Effects of multiscale porosity and pore interconnectivity on in vitro and in vivo degradation and biocompatibility of Fe–Mn–Cu scaffolds. J Mater Chem B 2021;9:4340–54. https://doi.org/10.1039/D1TB00641J.

[19] Drynda A, Hassel T, Bach FW, Peuster M. In vitro and in vivo corrosion properties of new iron–manganese alloys designed for cardiovascular applications. J Biomed Mater Res B Appl Biomater 2014;103:649–60. https://doi.org/10.1002/jbm.b.33234.

[20] Peuster M, Wohlsein P, Brügmann M, Ehlerding M, Seidler K, Fink C, et al. A novel approach to temporary stenting: degradable cardiovascular stents produced from corrodible metal—results 6–18 months after implantation into New Zealand white rabbits. Heart 2001;86:563–9. https://doi.org/10.1136/heart.86.5.563.

[21] Dargusch MS, Dehghan-Manshadi A, Shahbazi M, Venezuela J, Tran X, Song J, et al. Exploring the role of manganese on the microstructure, mechanical properties, biodegradability, and biocompatibility of porous iron-based scaffolds. ACS Biomater Sci Eng 2019;5:1686–702. https://doi.org/10.1021/acsbiomaterials.8b01497.

[22] Tonna C, Wang C, Mei D, Lamaka SV, Zheludkevich ML, Buhagiar J. Biodegradation behaviour of Fe-based alloys in Hanks' Balanced Salt Solutions: Part I. material characterisation and corrosion testing. Bioact Mater 2022;7:426–40. https://doi.org/10.1016/j.bioactmat.2021.05.048.

[23] Hermawan H, Purnama A, Dube D, Couet J, Mantovani D. Fe–Mn alloys for metallic biodegradable stents: degradation and cell viability studies. Acta Biomater 2010;6:1852–60. https://doi.org/10.1016/j.actbio.2009.11.025.

[24] Schinhammer M, Hänzi AC, Löffler JF, Uggowitzer PJ. Design strategy for biodegradable Fe-based alloys for medical applications. Acta Biomater 2010;6:1705–13. https://doi.org/10.1016/j.actbio.2009.07.039.

[25] Dehestani M, Trumble K, Wang H, Wang H, Stanciu LA. Effects of microstructure and heat treatment on mechanical properties and corrosion behavior of powder metallurgy derived Fe–30Mn alloy. Mater Sci Eng A 2017;703:214–26. https://doi.org/10.1016/j.msea.2017.07.054.

[26] Shuai C, Yang W, Yang Y, Pan H, He C, Qi F, et al. Selective laser melted Fe-Mn bone scaffold: microstructure, corrosion behavior and cell response. Mater Res Express 2019;7. https://doi.org/10.1088/2053-1591/ab62f5. 015404.

[27] Lin W, Qin L, Qi H, Zhang D, Zhang G, Gao R, et al. Long-term in vivo corrosion behavior, biocompatibility and bioresorption mechanism of a bioresorbable nitrided iron scaffold. Acta Biomater 2017;54:454–68. https://doi.org/10.1016/j.actbio.2017.03.020.

[28] Schinhammer M, Pecnik CM, Rechberger F, Hänzi AC, Löffler JF, Uggowitzer PJ. Recrystallization behavior, microstructure evolution and mechanical properties of biodegradable Fe–Mn–C(–Pd) TWIP alloys. Acta Mater 2012;60:2746–56. https://doi.org/10.1016/j.actamat.2012.01.041.

[29] Hufenbach J, Sander J, Kochta F, Pilz S, Voss A, Kühn U, et al. Effect of selective laser melting on microstructure, mechanical, and corrosion properties of biodegradable FeMnCS for implant applications. Adv Eng Mater 2020;22:2000182. https://doi.org/10.1002/adem.202000182.

[30] Niendorf T, Brenne F, Hoyer P, Schwarze D, Schaper M, Grothe R, et al. Processing of new materials by additive manufacturing: iron-based alloys containing silver for biomedical applications. Metall Mater Trans 2015;46:2829–33. https://doi.org/10.1007/s11661-015-2932-2.

[31] Bouaziz O, Allain S, Scott CP, Cugy P, Barbier D. High manganese austenitic twinning induced plasticity steels: a review of the microstructure properties relationships. Curr Opin Solid State Mater Sci 2011;15:141–68. https://doi.org/10.1016/j.cossms.2011.04.002.

[32] Schinhammer M, Gerber I, Hänzi AC, Uggowitzer PJ. On the cytocompatibility of biodegradable Fe-based alloys. Mater Sci Eng C 2013;33:782–9. https://doi.org/10.1016/j.msec.2012.11.002.

[33] Roth JA. Homeostatic and toxic mechanisms regulating manganese uptake, retention, and elimination. Biol Res 2006;39(1):47–57. https://doi.org/10.4067/s0716-97602006000100006.

[34] Kraus T, Moszner F, Fischerauer S, Fiedler M, Martinelli E, Eichler J, et al. Biodegradable Fe-based alloys for use in osteosynthesis: outcome of an in vivo study after 52 weeks. Acta Biomater 2014;10:3346–53. https://doi.org/10.1016/j.actbio.2014.04.007.

[35] Paul B, Lode A, Placht AM, Voß A, Pilz S, Wolff U, et al. Cell–Material interactions in direct contact culture of endothelial cells on biodegradable iron-based stents fabricated by laser powder bed fusion and impact of ion release. ACS Appl Mater Interfaces 2021;14:439–51. https://doi.org/10.1021/acsami.1c21901.

[36] Schinhammer M, Steiger P, Moszner F, Löffler JF, Uggowitzer PJ. Degradation performance of biodegradable Fe\Mn\C(\Pd) alloys. Mater Sci Eng C 2013;33:1882–93. https://doi.org/10.1016/j.msec.2012.10.013.

[37] Čapek J, Msallamová Š, Jablonská E, Lipovc J, Vojtěch D. A novel high-strength and highly corrosive biodegradable Fe-Pd alloy: structural, mechanical and in vitro corrosion and cytotoxicity study. Mater Sci Eng C 2017;79:550–62. https://doi.org/10.1016/j.msec.2017.05.100.

[38] Mandal S, Kishore V, Bose M, Nandi SK, Roy M. In vitro and in vivo degradability, biocompatibility and antimicrobial characteristics of Cu added iron-manganese alloy. J Mater Sci Technol 2021;84:159–72. https://doi.org/10.1016/j.jmst.2020.12.029.

[39] Huang T, Cheng J, Bian D, Zheng Y. Fe–Au and Fe–Ag composites as candidates for biodegradable stent materials. J Biomed Mater Res B Appl Biomater 2015;104:225–40. https://doi.org/10.1002/jbm.b.33389.

[40] Huang T, Cheng J, Zheng YF. In vitro degradation and biocompatibility of Fe–Pd and Fe–Pt composites fabricated by spark plasma sintering. Mater Sci Eng C 2014;35:43–53. https://doi.org/10.1016/j.msec.2013.10.023.

[41] Mijnendonckx K, et al. Antimicrobial silver: uses, toxicity and potential for resistance. Biometals 2013;26:609–21.

[42] Krüger JT, Hoyer KP, Filor V, Pramanik S, Kietzmann M, Meißner J, et al. Novel AgCa and AgCaLa alloys for Fe-based bioresorbable implants with adapted degradation. J Alloys Compd 2021;871:159544. https://doi.org/10.1016/j.jallcom.2021.159544.

[43] Hellmann M, Mehta SD, Bishai DM, Mears SC, Zenilman JM. The estimated magnitude and direct hospital costs of prosthetic joint infections in the United States, 1997 to 2004. J Arthroplasty 2009;25:766–71. https://doi.org/10.1016/j.arth.2009.05.025.

[44] Høiby N, Bjarnsholt T, Givskov M, Molin S, Ciofu O. Antibiotic resistance of bacterial biofilms. Int J Antimicrob Agents 2010;35:322–32. https://doi.org/10.1016/j.ijantimicag.2009.12.011.

[45] Wiesener M, Peters K, Taube A, Keller A, Hoyer K-P, Niendorf T, et al. Corrosion properties of bioresorbable FeMn-Ag alloys prepared by selective laser melting. Mater Corros 2017;68:1028–36. https://doi.org/10.1002/maco.201709478.

[46] Bagha PS, Khakbiz M, Sheibani S, Hermawan H. Design and characterization of nano and bimodal structured biodegradable Fe-Mn-Ag alloy with accelerated corrosion rate. J Alloys Compd 2018;767:955–65. https://doi.org/10.1016/j.jallcom.2018.07.206.

[47] Liu RY, He RG, Chen YX, Guo SF. Effect of Ag on the microstructure, mechanical and bio-corrosion properties of Fe–30Mn alloy. Acta Metall Sin 2019;32:1337–45. https://doi.org/10.1007/s40195-019-00911-5.

[48] Andreiev A, Hoyer KP, Grydin O, Frolov Y, Schaper M. Degradable silver-based alloys. Mater Werkst 2020;51:517–30. https://doi.org/10.1002/mawe.201900191.

[49] Krüger JT, Hoyer KP, Schaper M. Bioresorbable AgCe and AgCeLa alloys for adapted Fe-based implants. Mater Lett 2022;306:130890. https://doi.org/10.1016/j.matlet.2021.130890.

[50] Hoyer KP, Schaper M. Alloy design for biomedical applications in additive manufacturing. In: TMS 2019 148th annual meeting & exhibition supplemental proceedings. The minerals, metals & materials series. Cham: Springer; 2019. https://doi.org/10.1007/978-3-030-05861-6_44. The Minerals, Metals & Materials Series.

[51] Zhao JZ, Ahmed T, Jiang HX, He J, Sun Q. Solidification of immiscible alloys: a review. Acta Metall Sin 2017;30:1–28. https://doi.org/10.1007/s40195-016-0523-x.

[52] Wei C, Wang J, He Y, Li J, Beaugnon E. Solidification of immiscible alloys under high magnetic field: a review. Metals 2021;11:525. https://doi.org/10.3390/met11030525.

[53] ASM Alloy P D D. ASM alloy phase diagram database. 1999.

[54] Caligari Conti M, Mallia B, Sinagra E, Schembri Wismayer P, Buhagiar J, Vella D. The effect of alloying elements on the properties of pressed and non-pressed biodegradable Fe–Mn–Ag powder metallurgy alloys. Heliyon 2019;5:e02522. https://doi.org/10.1016/j.heliyon.2019.e02522.

[55] Sotoudehbagha P, Sheibanib S, Khakbiz M, Ebrahimi-Barough S, Hermawand H. Novel antibacterial biodegradable Fe-Mn-Ag alloys produced by mechanical alloying. Mater Sci Eng C 2018;88:88–94. https://doi.org/10.1016/j.msec.2018.03.005.

[56] Kraner J, Medved J, Godec M, Paulin I. Thermodynamic behavior of Fe-Mn and Fe-Mn-Ag powder mixtures during selective laser melting. Metals 2021;11:234. https://doi.org/10.3390/met11020234.

[57] Quan J, Lina K, Gua D. Selective laser melting of silver submicron powder modified 316L stainless steel: influence of silver addition on microstructures and performances. Powder Technol 2020;364:478–83. https://doi.org/10.1016/j.powtec.2020.01.082.

[58] Dadbakhsh S, Mertens R, Hao L, Van Humbeeck J, Kruth JP. Selective laser melting to manufacture "in situ" metal matrix composites: a review. Adv Eng Mater 2019;21:1801244. https://doi.org/10.1002/adem.201801244.

[59] Guo X, Zhang C, Tian Q, Yu D. Liquid metals dealloying as a general approach for the selective extraction of metals and the fabrication of nanoporous metals: a review. Mater Today Commun 2021;26:102007. https://doi.org/10.1016/j.mtcomm.2020.102007.

[60] McCue I, Gaskey B, Geslin PA, Karma A, Erlebacher J. Kinetics and morphological evolution of liquid metal dealloying. Acta Mater 2016;115:10–23. https://doi.org/10.1016/j.actamat.2016.05.032.

[61] Ferdous Z, Nemmar A. Health impact of silver nanoparticles: a review of the biodistribution and toxicity following various routes of exposure. Int J Mol Sci 2020;21. https://doi.org/10.3390/ijms21072375.

[62] Rodriguez-Contreras A, Punset M, Calero JA, Gil FJ, Ruperez E, Manero JM. Powder metallurgy with space holder for porous titanium implants: a review. J Mater Sci Technol 2021;76:129–49. https://doi.org/10.1016/j.jmst.2020.11.005.

[63] Yusop AH, Bakir AA, Shaharom NA, Abdul Kadir MR, Hermawan H. Porous biodegradable metals for hard tissue scaffolds: a review. Int J Biomater 2012;2012:641430. https://doi.org/10.1155/2012/641430.

[64] Khairallah SA, Anderson AT, Rubenchik A, King WE. Laser powder-bed fusion additive manufacturing: physics of complex melt flow and formation mechanisms of pores, spatter, and denudation zones. Acta Mater 2016;108:36–45. https://doi.org/10.1016/j.actamat.2016.02.014.

[65] Zhou X, Li K, Zhang D, Liu X, Ma J, Liu W, et al. Textures formed in a CoCrMo alloy by selective laser melting. J Alloys Compd 2015;631:153–64. https://doi.org/10.1016/j.jallcom.2015.01.096.

[66] Khairallah SA, Anderson A. Mesoscopic simulation model of selective laser melting of stainless steel powder. J Mater Process Technol 2014;214:2627–36. https://doi.org/10.1016/j.jmatprotec.2014.06.001.

[67] Papazoglou EL, Karkalos NE, Karmiris-Obratański P, Markopoulos AP. On the modeling and simulation of SLM and SLS for metal and polymer powders: a review. Arch Comput Methods Eng 2022;29:941–73. https://doi.org/10.1007/s11831-021-09601-x.

[68] Tan JHK, Sing SL, Yeong WY. Microstructure modelling for metallic additive manufacturing: a review. Virtual Phys Prototyp 2019;15:87–105. https://doi.org/10.1080/17452759.2019.1677345.

[69] Wang D, Wu S, Fu F, Mai S, Yang Y, Liu Y, et al. Mechanisms and characteristics of spatter generation in SLM processing and its effect on the properties. Mater Des 2017;117:121–30. https://doi.org/10.1016/j.matdes.2017.09.058.

[70] Qiu C, Panwisawas C, Ward M, Basoalto HC, Brooks JW, Attallah MM. On the role of melt flow into the surface structure and porosity development during selective laser melting. Acta Mater 2015;96:72–9. https://doi.org/10.1016/j.actamat.2015.06.004.

[71] Zhang G, Chen J, Zheng M, Yan Z, Lu X, Lin X, et al. Element vaporization of Ti-6Al-4V alloy during selective laser melting. Metals 2020;10(4):435. https://doi.org/10.3390/met10040435.

[72] Carluccio D, Demir AG, Caprio L, Previtali B, Bermingham MJ, Dargusch MS. The influence of laser processing parameters on the densification and surface morphology of pure Fe and Fe-35Mn scaffolds produced by selective laser melting. J Manuf Process 2019;40:113–21. https://doi.org/10.1016/j.jmapro.2019.03.018.

[73] Matthews MJ, Guss G, Khairallah SA, Rubenchik AM, Depond PJ, King WE. Denudation of metal powder layers in laser powder bed fusion processes. Acta Mater 2016;114:33—42. https://doi.org/10.1016/j.actamat.2016.05.017.

[74] Gunenthiram V, Peyre P, Schneider M, Dal M, Coste F, Koutiri I, et al. Experimental analysis of spatter generation and melt-pool behavior during the powder bed laser beam melting process. J Mater Process Technol 2017;251:376—86. https://doi.org/10.1016/j.jmatprotec.2017.08.012.

[75] Heeling T, Cloots M, Wegener K. Melt pool simulation for the evaluation of process parameters in selective laser melting. Addit Manuf 2017;14:116—25. https://doi.org/10.1016/j.addma.2017.02.003.

[76] Gunenthiram V, Peyre P, Schneider M, Dal M, Coste F, Fabbro R. Analysis of laser—melt pool—powder bed interaction during the selective laser melting of a stainless steel. J Laser Appl 2017;29. https://doi.org/10.2351/1.4983259. 022303.

[77] Andreau O, Koutiri I, Peyre P, Penot JD, Saintier N, Pessard E, et al. Texture control of 316L parts by modulation of the melt pool morphology in selective laser melting. J Mater Process Technol 2019;264:21—31. https://doi.org/10.1016/j.jmatprotec.2018.08.049.

[78] Bin Anwar A, Pham QC. Selective laser melting of AlSi10Mg: effects of scan direction, part placement and inert gas flow velocity on tensile strength. J Mater Process Technol 2017;240:388—96. https://doi.org/10.1016/j.jmatprotec.2016.10.015.

[79] Panwisawas C, Qiu C, Anderson MJ, Sovani Y, Turner RP, Attallah MM, et al. Mesoscale modelling of selective laser melting: thermal fluid dynamics and microstructural evolution. Comput Mater Sci 2017;126:479—90. https://doi.org/10.1016/j.commatsci.2016.10.011.

[80] Wu YC, San CH, Chang CH, Lin HJ, Marwan R, Baba S, et al. Numerical modeling of melt-pool behavior in selective laser melting with random powder distribution and experimental validation. J Mater Process Technol 2018;254:72—8. https://doi.org/10.1016/j.jmatprotec.2017.11.032.

[81] Patel S, Vlasea M. Melting modes in laser powder bed fusion. Materialia 2020;9:100591. https://doi.org/10.1016/j.mtla.2020.100591.

[82] King WE, Barth HD, Castillo VM, Gallegos GF, Gibbs JW, Hahn DE, et al. Observation of keyhole-mode laser melting in laser powder-bed fusion additive manufacturing. J Mater Process Technol 2014;214:2915—25. https://doi.org/10.1016/j.jmatprotec.2014.06.005.

[83] Sing NB, Mostavan A, Hamzah E, Mantovani D, Hermawan H. Degradation behavior of biodegradable Fe35Mn alloy stents. J Biomed Mater Res B Appl Biomater 2014;103:572—7. https://doi.org/10.1002/jbm.b.33242.

[84] Scarcello E, Lison D. Are Fe-based stenting materials biocompatible? A critical review of in vitro and in vivo studies. J Funct Biomater 2020;11. https://doi.org/10.3390/jfb11010002.

[85] Loffredo S, Paternoster C, Giguère N, Vedani M, Mantovani D. Effect of silver on corrosion behavior of plastically deformed twinning-induced plasticity steel for biodegradable stents. J Miner Met Mater Soc 2020;72:1892—901. https://doi.org/10.1007/s11837-020-04111-w.

[86] Chiang WC, Tseng I-S, Møller P, Rischel Hilbert L, Tolker Nielsen T, Wu JK. Influence of silver additions to type 316 stainless steels on bacterial inhibition, mechanical properties, and corrosion resistance. Mater Chem Phys 2010;119:123—30. https://doi.org/10.1016/j.matchemphys.2009.08.035.

[87] Gosheger G, Hardes J, Ahrens H, Streitburger A, Buerger H, Erren M, et al. Silver-coated megaendoprostheses in a rabbit model—an analysis of the infection rate and toxicological side effects. Biomaterials 2004;25:5547—56. https://doi.org/10.1016/j.biomaterials.2004.01.008.

[88] Le T-N, Lo Y-L. Effects of sulfur concentration and Marangoni convection on melt-pool formation in transition mode of selective laser melting process. Mater Des 2019;179:107866. https://doi.org/10.1016/j.matdes.2019.107866.

[89] Lebba M, Astarita A, Mistretta D, Colonna I, Liberini M, Scherillo F, et al. Influence of powder characteristics on formation of porosity in additive manufacturing of Ti-6Al-4V components. J Mater Eng Perform 2017;26:4138—47. https://doi.org/10.1007/s11665-017-2796-2.

[90] Gürtler FJ, Karg M, Dobler M, Kohl S, Tzivilsky I, Schmidt M. Influence of powder distribution on process stability in laser beam melting: analysis of melt pool dynamics by numerical simulations. 25th Solid Freeform Fabrication Symposium 2014;25:1099—117.

[91] Ghods S, Schur R, Schultz E, Pahuja R, Montelione A, Wisdom C, et al. Powder reuse and its contribution to porosity in additive manufacturing of Ti6Al4V. Materialia 2021;15:100992. https://doi.org/10.1016/j.mtla.2020.100992.

[92] Choo H, Shama KL, Bohling J, Ngo A, Xiao X, Ren Y, et al. Effect of laser power on defect, texture, and microstructure of a laser powder bed fusion processed 316L stainless steel. Mater Des 2019;164:107534. https://doi.org/10.1016/j.matdes.2018.12.006.

[93] Ng GKL, Jarfors AEW, Bi G, Zheng HY. Porosity formation and gas bubble retention in laser metal deposition. Appl Phys A 2009;97:641—9. https://doi.org/10.1007/s00339-009-5266-3.

[94] Chen H, Yan W. Spattering and denudation in laser powder bed fusion process: multiphase flow modelling. Acta Mater 2020;196:154—67. https://doi.org/10.1016/j.actamat.2020.06.033.

[95] Ahsan RU, Cheepu M, Ashiri R, Kim TH, Jeong C, Park YD. Mechanisms of weld pool flow and slag formation location in cold metal transfer (CMT) gas metal arc welding (GMAW). Weld World 2017;61:1275—85. https://doi.org/10.1007/s40194-017-0489-y.

[96] Ogino K, Hara S, Miwa T, Kimoto S. The effect of oxygen content in molten iron on the interfacial tension between molten iron and slag. Trans Iron Steel Inst Japan 1984;24:522—31. https://doi.org/10.2355/isijinternational1966.24.522.

[97] Morrow BM, Lienert TJ, Knapp CM, Sutton JO, Brand MJ, Pacheco RM, et al. Impact of defects in powder feedstock materials on microstructure of 304L and 316L stainless steel produced by additive manufacturing. Metall Mater Trans 2018;49:3637—50. https://doi.org/10.1007/s11661-018-4661-9.

[98] Shannon G, White L, Sridhar S. Modeling inclusion approach to the steel/slag interface. Mater Sci Eng A 2008;495:310—5. https://doi.org/10.1016/j.msea.2007.09.087.

[99] Zhong Y, Liu L, Wikman S, Cui D, Shen Z. Intragranular cellular segregation network structure strengthening 316L stainless steel prepared by selective laser melting. J Nucl Mater 2016;470:170—8. https://doi.org/10.1016/j.jnucmat.2015.12.034.

[100] Acharya R, Sharon JA, Staroselsky A. Prediction of microstructure in laser powder bed fusion process. Acta Mater 2017;124:360—71. https://doi.org/10.1016/j.actamat.2016.11.018.

[101] Wan HY, Zhou ZJ, Li CP, Chen GF, Zhang GP. Effect of scanning strategy on grain structure and crystallographic texture of Inconel 718 processed by selective laser melting. J

Mater Sci Technol 2018;34:1799–804. https://doi.org/10.1016/j.jmst.2018.02.002.

[102] Thijs L, Verhaeghe F, Craeghs T, Van Humbeeck J, Kruth JP. A study of the microstructural evolution during selective laser melting of Ti–6Al–4V. Acta Mater 2010;58:3303–12. https://doi.org/10.1016/j.actamat.2010.02.004.

[103] Zhang X, Liu B, Zhou X, Wang J, Hashimoto T, Luo C, et al. Laser welding introduced segregation and its influence on the corrosion behaviour of Al-Cu-Li alloy. Corrosion Sci 2018;135:177–91. https://doi.org/10.1016/j.corsci.2018.02.044.

[104] Yadroitsev I, Krakhmalev P, Yadroitsava I, Johansson S, Smurov I. Energy input effect on morphology and microstructure of selective laser melting single track from metallic powder. J Mater Process Technol 2013;213:606–13. https://doi.org/10.1016/j.jmatprotec.2012.11.014.

[105] Yadroitsev G, Krakhmalev P, Yadroitsava I, Johansson S, Smurov I. Energy input effect on morphology and microstructure of selective laser melting single track from metallic powder. Opt Laser Technol 2022;147:107621. https://doi.org/10.1016/j.optlastec.2021.107621.

[106] Houdremont E. Einführung in die Sonderstahlkunde. Berlin: Springer; 1935.

[107] Nadammal N, Cabeza S, Mishurova T, Thiede T, Kromma A, Seyfert C, et al. Effect of hatch length on the development of microstructure, texture and residual stresses in selective laser melted superalloy Inconel 718. Mater Des 2017;134:139–50. https://doi.org/10.1016/j.matdes.2017.08.049.

[108] Thijs L, Montero Sistiaga ML, Wauthle R, Xie Q, Kruth JP, Van Humbeeck J. Strong morphological and crystallographic texture and resulting yield strength anisotropy in selective laser melted tantalum. Acta Mater 2013;61:4657–68. https://doi.org/10.1016/j.actamat.2013.04.036.

[109] Woolf PL, Zellars GR, Foerster E, Morris JP. Vapor pressures of liquid manganese and liquid silver. United States Department Of The Interior, Bureau Of Mines; 1960.

[110] Wei K, Wang Z, Zeng X. Influence of element vaporization on formability, composition, microstructure, and mechanical performance of the selective laser melted Mg–Zn–Zr components. Mater Lett 2015;156:187–90. https://doi.org/10.1016/j.matlet.2015.05.074.

[111] Ferguson FT, Nuth JA, Johnson NM. Thermogravimetric measurement of the vapor pressure of iron from 1573 K to 1973 K. J Chem Eng Data 2004;49:497–501. https://doi.org/10.1021/je034152w.

[112] Sárosi Z, Knapp W, Kunz A, Wegener K. Evaluation of reflectivity of metal parts by a thermo-camera. InfraMation 2010 Proceedings 2010:475–86. https://doi.org/10.3929/ETHZ-A-006206911.

[113] Arjunan A, Robinson J, Al Ani E, Heaselgrave W, Baroutaji A, Wang C. Mechanical performance of additively manufactured pure silver antibacterial bone scaffolds. J Mech Behav Biomed Mater 2020;112:104090. https://doi.org/10.1016/j.jmbbm.2020.104090.

[114] Xiong W, Hao L, Li Y, Tang D, Cui Q, Feng Z, et al. Effect of selective laser melting parameters on morphology, microstructure, densification and mechanical properties of supersaturated silver alloy. Mater Des 2019;170:107697. https://doi.org/10.1016/j.matdes.2019.107697.

[115] Dietrich K, Jaensson N, Buttinoni I, Volpe G, Isa L. Microscale marangoni surfers. Phys Rev Lett 2020;125. https://doi.org/10.1103/PhysRevLett.125.098001. 098001.

[116] Panwisawas C, Qiu CL, Sovani Y, Brooks JW, Attallah MM, Basoalto HC. On the role of thermal fluid dynamics into the evolution of porosity during selective laser melting. Scripta Mater 2015;105:14–7. https://doi.org/10.1016/j.scriptamat.2015.04.016.

[117] Panwisawas C, Sovani Y, Turner RP, Brooks JW, Basoalto HC, Choquet I. Modelling of thermal fluid dynamics for fusion welding. J Mater Process Technol 2018;252:176–82. https://doi.org/10.1016/j.jmatprotec.2017.09.019.

[118] Li T, Song G, Yu P, Liu L. Interfacial microstructure evolution in fusion welding of immiscible Mg/Fe system. Mater Des 2019;181:107903. https://doi.org/10.1016/j.matdes.2019.107903.

[119] Egry I, Ricci E, Novakovic R, Ozawa S. Surface tension of liquid metals and alloys — recent developments. Adv Colloid Interface Sci 2010;159:198–212. https://doi.org/10.1016/j.cis.2010.06.009.

[120] Hibiya T, Ozawa S. Effect of oxygen partial pressure on the marangoni flow of molten metals. Cryst Res Technol 2013;48:208–13. https://doi.org/10.1002/crat.201200514.

[121] Murarka RN, Lu WK, Hamielec AE. Effect of dissolved oxygen on the surface tension of liquid iron. Can Metall Quart 1975;14:111–5. https://doi.org/10.1179/000844375795050337.

[122] Ozawa S, Takahashi S, Suzuki S, Sugawara H, Fukuyama H. Relationship of surface tension, oxygen partial pressure and temperature for molten iron. Jpn J Appl Phys 2011;50:11RD05. https://doi.org/10.1143/JJAP.50.11RD05.

[123] Ozawa S, Morohoshi K, Hibiya T, Fukuyama H. Influence of oxygen partial pressure on surface tension and its temperature coefficient of molten iron. J Appl Phys 2011;109. https://doi.org/10.1063/1.3527917. 014902.

[124] Hoai LT, Lee J. Effect of surface adsorption of carbon on the surface tension of liquid Fe–Mn–C alloys. J Mater Sci 2012;47:8303–7. https://doi.org/10.1007/s10853-012-6836-x.

[125] Ozawa S, Morohoshi K, Hibiya T, Fukuyama H. Influence of oxygen partial pressure on surface tension of molten silver. J Appl Phys 2010;107. https://doi.org/10.1063/1.3275047. 014910.

[126] Louvis E, Fox P, Sutcliffe CJ. Selective laser melting of aluminium components. J Mater Process Technol 2011;211:275–84. https://doi.org/10.1016/j.jmatprotec.2010.09.019.

[127] C. Thelwlis, D.R. Milner, Inclusion formation in arc welding, Welding Research Supplement 281-s (1977).

[128] Teng Q, Li S, Wei Q, Shi Y. Investigation on the influence of heat treatment on Inconel 718 fabricated by selective laser melting: microstructure and high temperature tensile property. J Manuf Process 2021;61:35–45. https://doi.org/10.1016/j.jmapro.2020.11.002.

[129] Baroud CN. Thermocapillarity. In: Li D, editor. Encyclopedia of microfluidics and nanofluidics. Boston: Springer; 2013.

[130] Zhang C, Gao M, Wang D, Yin J, Zeng X. Relationship between pool characteristic and weld porosity in laser arc hybrid welding of AA6082 aluminum alloy. J Mater Process Technol 2017;240:217–22. https://doi.org/10.1016/j.jmatprotec.2016.10.001.

[131] Zhao Y, Aoyagi K, Yamanaka K, Chiba A. Role of operating and environmental conditions in determining molten pool dynamics during electron beam melting and selective laser melting. Addit Manuf 2020;36:101559. https://doi.org/10.1016/j.addma.2020.101559.

 *crystals*

Article

Adjustment of AgCaLa Phases in a FeMn Matrix via LBM for Implants with Adapted Degradation

Jan Tobias Krüger [1,2]

[1] Materials Science, Paderborn University, Mersinweg 7, 33100 Paderborn, Germany; krueger@lwk.upb.de
[2] DMRC-Direct Manufacturing Research Center, Paderborn University, Mersinweg 3, 33100 Paderborn, Germany

Abstract: For many applications, implants overtake body function for a certain time. Bioresorbable implants reduce patient burden as they prevent adverse consequences due to remaining implants or operations for removal. Such materials are in clinical use but do not fulfill the requirements of all applications. Iron (Fe) is promising to develop further bioresorbable materials as it offers biocompatibility and good mechanical properties. Alloying, e.g., with manganese (Mn), is necessary to adapt the mechanical behavior and the degradation rate. However, the degradation rate of FeMn is too low. The creation of phases with high electrochemical potential evokes anodic dissolution of the FeMn, increasing the degradation rate. Therefore, silver (Ag), which is insoluble with Fe, has high potential, is biocompatible, and offers antibacterial properties, can be used. Powder-based processes such as laser beam melting (LBM) are favorable to process such immiscible materials. A degradable Ag alloy has to be used to enable the dissolution of Ag phases after the FeMn. This study reports first about the successful processing of FeMn with 5 wt.% of a degradable Ag–calcium–lanthanum (AgCaLa) alloy and enables further targeted adaption due to the gained understanding of the effects influencing the morphology and the chemical composition of the Ag phases.

Keywords: silver alloys; corrosion; biomedical application; laser beam melting; bioresorbable metal; iron alloys

Citation: Krüger, J.T. Adjustment of AgCaLa Phases in a FeMn Matrix via LBM for Implants with Adapted Degradation. *Crystals* **2022**, *12*, 1146. https://doi.org/10.3390/cryst12081146

Academic Editors: Bolv Xiao and Kay-Peter Hoyer

Received: 26 July 2022
Accepted: 12 August 2022
Published: 15 August 2022

Publisher's Note: MDPI stays neutral with regard to jurisdictional claims in published maps and institutional affiliations.

1. Introduction

To enable the treatment of diseases and injuries with implants, suitable implants with properties adapted for each application are necessary. This forms a need for new materials with adapted properties. Thus, new materials are developed to improve the quality of patients' lives and reduce patients' burdens. [1–3] All materials for implants need to be biocompatible, which describes the absence of any adverse consequences including inflammatory, toxicity, carcinogenic properties, etc. The most currently used materials guarantee biocompatibility due to inert material behavior. However, the application of not inert materials enables the utilization of beneficial features of materials for implants, such as the antibacterial effect of released ions [1,2,4,5]. In the actual research e. g. bioresorbable materials to prevent adverse consequences by remaining implants and operations for removal of implants or materials with antibacterial properties preventing implant infections are addressed [3,5–7].

Magnesium (Mg)-based materials or polymers are already applied successfully as degradable implants [1,5,8,9], but currently available materials are not suited for all possible applications because of reduced mechanical properties compared to conventional iron (Fe)-based implant materials [2,8,10]. Thus, further degradable materials need to be developed. Due to good mechanical properties and biocompatibility, Fe alloys are promising as degradable materials. Hence, the presented investigation aimed to develop a bioresorbable Fe-based material with antibacterial properties [5,8,11–13]. Alloying of Fe is required to obtain suited mechanical properties and adapted degradation behavior, as pure Fe degrades too slowly [3,5,14,15]. Manganese improves the mechanical properties due to its twinning-induced plasticity (TWIP) and degradation rate, as it reduces the electrochemical potential. In addition, segregations of Mn lead to differences in the local electrochemical potential. As a material with lower electrochemical potential is increasingly attacked by corrosion, local differences improve the anodic dissolution of areas with low potential [12,15–19]. For such FeMn alloys, promising results regarding biocompatibility have been obtained [16,20–24].

Probably due to the formation of degradation inhibiting layers, the degradation rate of FeMn is for nearly all applications too low and the material has to be further adapted [13,20,22,25]. The creation of targeted noble phases within a FeMn matrix is an approach resulting in the increased anodic dissolution of the FeMn matrix [25–28]. When two metals with different electrochemical potentials are in contact and an electrolyte is present, the atoms of the less noble material are more likely to be dissolved than when without contact with a more noble material. Due to their high electrochemical potential, noble elements are well suited to create noble phases. In particular, Ag is promising to form noble phases within FeMn as it is insoluble with Fe, biocompatible, and has antibacterial properties [26,29–31]. This immiscibility enables the existence of noble phases in an unchanged FeMn matrix, which makes it well suited apart from its low degradation rate. Enhanced degradation due to the addition of Ag is reported by a few studies [12,26,29,32,33]. The addition of 5 wt.% Ag was selected for the presented investigations as it enables suited properties regarding mechanical behavior, biocompatibility, and degradation behavior [26,29,34,35]. However, this is a guess and it might be necessary to adapt the amount of Ag, based on further investigations of the material, but 5 wt.% is promising. Furthermore, the Ag has to be adapted by alloying as it would remain after degradation of the matrix and could cause adverse consequences. For example, the Ag particles could be transported to sensitive organs in the human body such as the brain or the lung [36–38]. Thus, biocompatible and degradable Ag alloys are required to substitute for the pure Ag. The electrochemical potential of these alloys is still high enough to evoke anodic dissolution of the FeMn, but they enable the dissolution of the Ag phases after the matrix material [31,39–41]. In the presented investigations a degradable AgCaLa alloy with 12 wt.% Ca and 5 wt.% La, developed by Krüger et al. [31], was utilized.

For the successful realization of the FeMn–AgCaLa material concept, dispersed Ag-CaLa phases in the FeMn matrix have to be evoked. To process such immiscible materials and obtain an adapted morphology of Ag phases is challenging [42,43]. Powder-based metallurgical processes such as laser beam melting enable the processing of such materials, as mechanically mixed powders can be used [29,44,45]. Due to the small melt pool, strong melt flow, as well as rapid melting and solidification, homogenous and adaptable Ag phases in an iron-based matrix can be obtained via LBM [26,35,46,47]. Apart from the morphology of the Ag phases, the chemical composition has to be addressed. To achieve the required chemical composition might be a concern, as between immiscible materials distinctive diffusion of alloying elements might occur [35,48]. Thus LBM is a promising technique, as the short time at high temperatures reduces the time for diffusion. However, even for the processing of FeMn with pure Ag via LBM, distinctive diffusion of Mn from the matrix to the Ag is observed [35]. Thus, an adaption of the LBM strategy has to address a reduced diffusion of alloying elements as well as the morphology of the Ag phases since the chemical composition of Ag phases is important for the electrochemical potential and the degradation performance [31,39–41]. LBM strategies with reduced energy input are favorable as they reduce the time at high temperatures, but for these strategies, a certain porosity has to be accepted [35].

However, the uptake of Ca and La by the FeMn matrix is not critical as both elements can be used as deoxidizers [49,50]. Thus, diffusion of Ca and La is acceptable as long as the Ag phases remain degradable. A diffusion of Mn from Fe into Ag reduces the Mn content within the matrix material, whereby the matrix would keep the austenitic structure and the TWIP effect even if the Mn content is reduced by 2 wt.% [19]. On the one hand, Mn might contribute to the degradation of the Ag phases after the dissolution of the matrix, but on the other hand, the uptake of Mn reduces the electrochemical potential of pure Ag [35]. In summary, the diffusion of alloying elements is noncritical for the matrix material. The same can be assumed for the AgCaLa alloy. For this alloy, an impact on the degradability and electrochemical potential will also be present, but this should be acceptable since the diffusion is not too pronounced.

Furthermore, the mechanical properties are decisive for the intended application. Due to its immiscibility, the inclusions of Ag do not have a direct connection to the FeMn and act like pores concerning the load capability, but the mechanical properties are still satisfying as the ductile character of FeMn enables the reduction of tension peaks [26,35]. Strategies are being investigated to obtain a bone-like stiffness via targeted porosity to avoid stress shielding; thus, the Ag phases might address this concern by the way [51,52].

However, to develop a suited material, the Ag phases need to be adapted and, therefore, knowledge of the interaction of both melts throughout the LBM process, respectively, and the mechanisms responsible for the generation of the Ag inclusions are mandatory to enable a targeted adaption. Gaining knowledge regarding the influencing physical effects during LBM to enable the targeted adaption is the focal point of current research [53–56]. As the physical effects cannot be observed directly, the resulting microstructure and properties of the processed material are investigated, knowledge concerning the physical effects is transferred, and the melt pools are observed to generate models of the LBM process [53,57–59].

The laser radiation causes the continuous formation of bulk material due to sequential proceeding sub-processes. The most decisive sub-processes depending on further physical effects and interactions with each other are the absorption of laser radiation, melting and consolidation, melt flow, heat dissipation, and solidification [53,55,57,59]. Further effects such as spattering or chemical reactions are caused by the sub-processes [55,57,60,61], whereby, e.g., the adsorption depends not only on the absorptivity of the solid and liquid material as an eventually formed keyhole, reflection within the powder bed, and deflection by the vapor plume, etc., but influence the energy intake, too [54,62,63]. A depression is formed if the recoil pressure of vaporized material is high enough to displace molten material against surface tension [64–66]. As the laser is reflected within a depression it further increases energy absorption and more material evaporates. Thus, the depression supports itself and once a threshold value of energy input is exceeded, a keyhole is formed [64,67]. Due to increased energy absorption in the so-called keyhole welding mode, the size of the melt pool can increase [64,68]. As the addition of alloying elements with high vapor pressure supports the formation of a keyhole, they can lower the energy input needed for the keyhole welding mode [65–67].

Furthermore, the spattering influences the final structure and composition of the mixed material as spattering depends on the material properties as well as process characteristics such as energy density [57,60,69,70]. Ejections of liquid material mainly result from recoil pressure and Marangoni forces, respectively, and instabilities of the melt flow [60,71], whereby a keyhole significantly contributes to spattering [57,60,70]. The intensity of effects causing spattering depends, e.g., on the viscosity of the melt and the surface tension [54,57,60,71]. Aside from the ejection of molten material, the non-molten powder is ejected due to an upward gas flow resulting from heating and vaporization in the laser spot. Additionally, the material-dependent evaporation can cause a bow-off impulse, accelerating powder particles from the surrounding powder bed [57,70,72].

Aside from the physical effects, the chemical properties of the reactivity of the processed material influence the LBM process, as elements with high reactivity can react with each other or with the remaining oxygen within the building chamber [73–75]. As the reactivity increases with temperature, the chemical reactions can be so distinctive that developed oxides deteriorate the connection between adjacent and superimposed melt tracks [73,74,76,77]. Additionally, slag positioned on the top can be incorporated into the bulk material, deteriorating the mechanical properties [74,78,79].

Thus, manifold interacting physical effects depending on the material properties and process conditions have to be considered to develop a model that explains the continuous melting process and structure of the manufactured parts. Therefore, at first, the main influencing variables, which are the surface tension and the recoil pressure, have to be incorporated. Following this, further effects can be integrated to improve the model [53,54,57]. However, the evaporation is decisive for the welding mode as well as consequently the appearance of the melt pool. Thus, via the shape of the melt pool, the welding mode can be identified. Flat melt pools are characteristic of the conduction welding mode with heat transfer mainly by conduction. Deep and narrow melt pools arise from the before-mentioned keyhole welding mode with heat transferred by convection. A transition mode with proportions of both modes is defined between the other modes [53,80,81].

Based on the before-mentioned knowledge from the literature, and by the results presented in this study, a model was developed to explain the interaction of both melts and enable targeted adaption. As this model is based on the results gained for processing of FeMn with pure Ag by Krüger et al. [35], the main statements relevant to the present study are briefly summarized as following, beginning with separation of a melt pool into three sections where different physical effects are decisive [53]. The first part of the melt pool is predominantly influenced by the recoil pressure causing depression, which provokes the transition welding mode and melt flow due to displacement. As the recoil effect acts similarly regarding FeMn and pure Ag, it causes the emergence of Ag into the melt pool, which supports the diffusion of alloying elements from one metal into another metal.

For the morphology of the Ag phases, the tail end and transition section determined by surface tension are decisive, since during solidification the position of Ag is defined and the location in the moment of solidification is defined by the former melt flow. The surface tension leads to the accumulation of Ag on the top as this arrangement requires the lowest surface energy. Since accumulation reduces the surface, it minimizes the surface energy. Due to the lower surface tension of Ag compared to FeMn, a surface of Ag on the top requires less energy than a surface of FeMn on the top would require. Thus, once Ag reaches the top surface it is trapped there. The Ag is especially positioned at the flanks of the melt pools as an outward-directed vortex because the Marangoni force transports Ag to the melt pool sides.

2. Experimental Procedures

The production, preparation, and investigation of the samples were almost identical to the methods applied in Krüger et al. [35]. Thus, the results of both studies are well comparable.

Argon-gas-atomized FeMn powder (nominal composition: 22 wt.% Mn, 0.6 wt.% C, Fe bal.; Nanoval GmbH & Co. KG, Berlin, Germany) and AgCaLa powder (Nanoval GmbH & Co. KG, Berlin, Germany) were used in this study. To measure the chemical composition of the AgCaLa powder, inductive coupled optical emission spectroscopy (ICP-OES) was deployed. Feedstock for the AgCaLa powder was an alloy with 12.9 wt.% Ca and 5.7 wt.% La (ICP-OES). The particle size distribution (PSD) was determined with a Mastersizer 2000 (Malvern Panalytical GmbH, Kassel, Germany). An HL-2000-HP-FHSA (Ocean Optics, Inc., Orlando, FL, USA) was utilized to determine the absorption of the powder bed.

The FeMn was mixed with 5 wt.% of AgCaLa with a drum hoop mixer (30 min at 60 rpm) using half-filled powder containers. To guarantee a humidity of less than 5%, the powder was vacuum dried. With an SLM 280^{HL} 2.0 (SLM Solutions Group AG, Lübeck, Germany) the powder was processed in an inert gas atmosphere (argon (4.6), oxygen level < 0.03%). Based on previously unpublished work, the parameters were selected (Table 1). All samples (cubes: $10 \times 10 \times 10$ mm^3, cylinders for computer tomographic (CT) analysis: $\varnothing\, 5 \times 10$ mm^3, blocks for tensile specimens: $13 \times 36 \times 10$ mm^3) were produced in one built-job. According to the density and the achieved AgCaLa phase content, the samples were designated as "HD" for high density, "LD" for low density, and "LSC" for samples with low silver content.

Table 1. Parameters conducted for processing via LBM.

Strategy	HD	LD	LD-2	LSC	LSC-2
Slicing in μm	50	50	50	50	50
Scanning velocity in mm/s	850	900	900	700	700
Laser power in W	260	220	220	300	280
Hatching in μm	150	190	130	110	110

The scanning direction was rotated 67° between the following layers. To reveal an idea of the melt pool appearance, some samples without rotation were fabricated as well. These samples are indicated by a "-0" at the end of the enumeration, e.g., HD-0.

Miniature dog-bone specimens were prepared for the tensile tests (measuring length: 8 mm, width: 3 mm, thickness: 1.5 mm, radius: 2 mm). The shapes of the specimens followed DIN EN ISO 6892-6 but were not in accordance with the criteria of proportional specimens. The specimens were eroded out of bulk material (HD) manufactured horizontally aligned and ground up to the final step P2500 with silicon carbide (SiC) paper.

Grinding up to P4000 was conducted iteratively for the preparation for the microstructural investigations. To avoid the impact of edge effects, 1 mm or more material was removed. A mixture of 50% corrosion inhibiting lubricant (CUTLUB (diluted, 3%), Cloeren Technology GmbH, Wegberg, Germany) and silica polishing suspension (grit size: 50 nm, Cloeren Technology GmbH, Wegberg, Germany) was used for final vibration polishing for 12 h. Selected samples were etched with 0.5% HNO_3 acid from 30 s up to 150 s.

An Axiophot (Carl Zeiss AG, Oberkochen, Germany) and a Keyence VHX5000 digital microscope (KEYENCE Corporation, Osaka, Japan) were applied for light microscopy (LM). A Zeiss Ultra Plus (Carl Zeiss AG, Oberkochen, Germany) field emission scanning electron microscope (FE-SEM) with secondary electron (SE) and backscatter electron (BSE) detectors was used for further investigation. The electron backscatter diffraction (EBSD) detector DIGIVIEW 5 (AMETEK, Berwyn, PA, USA) enabled the investigation of the grain structure, whereas the chemical composition was detected with an energy-dispersive X-ray spectroscope (EDS) detector Octane Pro (AMETEK, Berwyn, PA, USA). All EBSD-data presented in the Results Section were not cleaned. The micro-CT SkyScan 1275 (Bruker Corporation, Billerica, MA, USA) enabled the three-dimensional observation of the structure of the Ag phases. Via wavelength dispersive X-ray analysis (WDX) with a Jeol JXA-8100 (Jeol Denshi K.K., Akishima, Tokyo, Japan), the content of elements of the AgCaLa within the matrix was determined.

An MTS 858 Table Top System (MTS Systems Corporation, Eden Prairie, MN, USA) was used to perform quasi-static tensile tests with a testing velocity of 0.13 mm/min (strain rate less than 0.0025 1/s). The tests were performed following DIN EN ISO 6892-6 and an extensometer was used.

3. Results and Discussion

As the feedstock properties are crucial for the LBM process, the powder used was characterized in a first step. Both powders had a suited particle size distribution (PSD) (Figure 1a); the morphology of the AgCaLa powder was appropriate as it consisted mainly of spherical particles with only a few satellites (Figure 1d). The same applied to the FeMn powder [35]. The AgCaLa powder feedstock contained 12.4 wt.% Ca, 5 wt.% La, 0.3 wt.% O, and Ag bal. measured via ICP-OES. Thus, the chemical composition met the intended content of alloying elements (12 wt.% Ca, 5 wt.% La) and the powder took up only a little O during gas atomization despite the high reactivity of Ca and La. The surface structure of the AgCaLa particles confirmed the achievement of a capable AgCaLa feedstock, as a lamellar structure was present, as observed for the conventionally cast alloy [31]. The addition of AgCaLa influenced the absorption of the laser radiation, as for FeMn adsorption of 69.3% was determined in contrast to 61.8% for FeMn with AgCaLa. However, sufficient adsorption can still be expected. Thus, the feedstock was well-suited to obtain a material with the intended properties.

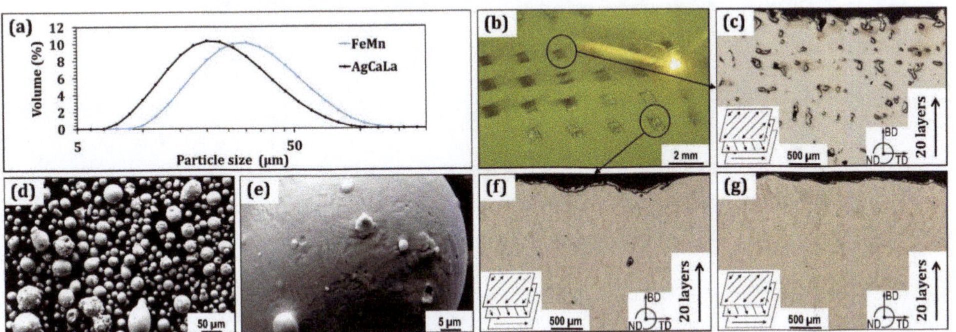

Figure 1. (**a**) PSD of powder feedstock; (**d**,**e**) SEM-SE: gas-atomized AgCaLa particles; (**b**) processing via LBM with different parameter settings; (**c**,**f**,**g**) LM: cross-section of samples with varying AgCaLa content in the bulk material process strategies: (**c**) LD-2, (**f**) LSC-2, (**g**) LSC.

Since the process parameters significantly influence the behavior of the AgCaLa during the LBM process, differences in smoke evolution, sample top surfaces, and bulk material were observed. Some samples comprised metallic blanc as-built top surfaces while others were matt (Figure 1b). This matt surface might be due to the formation of slag, as well as the deposition of welding spatter. This hypothesis is substantiated by the increased deposition of smoke marks near the samples with matt surfaces. Cross-sections confirmed differences, as the top surface of matt samples was significantly rougher (Figure 1c,f,g). On top of all samples, some AgCaLa phases were deposited, but the AgCaLa phases were more distinctive for samples with matt surfaces. However, the distribution of AgCaLa in the bulk material differed more significantly as in some samples nearly no AgCaLa was incorporated, whereas other samples contained well-dispersed AgCaLa phases connected to some porosity. The lack of AgCaLa in some of the samples was not caused by inhomogeneous deposition of the powder feedstock, as the AgCaLa depositions on the top of these samples proved. Additionally, the good distribution of both components in the powder bed was confirmed by investigations of the chemical distribution [35]. Thus, the incorporation of AgCaLa into the bulk material strongly depended on the process parameters. This can be utilized to manufacture material with a graded distribution of AgCaLa, as demonstrated further in Figure 5. The settings used for the sample shown in Figure 1g were applied to obtain areas without AgCaLa.

As a surface like the final top surface was remelted with the manufacturing of each following layer, the upper surface was characterized by its morphology as well as its chemical composition. The as-built top surface was enriched with Ag and depleted of Fe (Figure 2a), confirming the formation of an AgCaLa layer on the top of samples containing AgCaLa in the bulk material, whereby some of the areas with a high Fe content were due to non-molten particles adhering to the top surface. These particles are transported to the melt pool behind the interaction zone with the laser by the inward gas flow [60–62,82]. Consequently, these powder adhesions do not originate out of the melt pool, and the AgCaLa layer formed by the melting process is more distinctive.

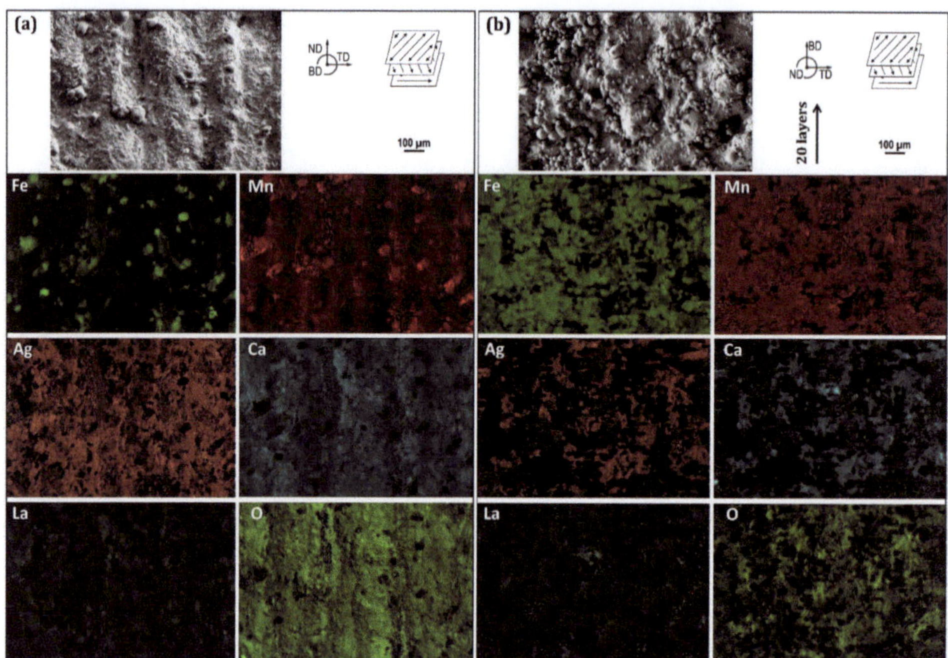

Figure 2. Corresponding SEM-SE and SEM-EDS of as-built surfaces (HD): (**a**) top surface with enrichment of Ag and O, (**b**) side surface.

The inhomogeneous distribution of Ca, Mn, and La together with O on the top surface indicates the formation of slag on the top surface containing various chemical compounds of the alloying elements and O. This surface composition explains the observed matt surface appearance and smoke evolution (Figure 1b). Thus, chemical reactions between alloying elements and remaining oxygen occur during the LBM process.

The side surface was characterized by adhering powder particles from the adjacent powder bed and enrichment of Ag, as well as an inhomogeneous distribution of the alloying elements and O, like on the top surface, but less distinctive. The adhering FeMn particles covered some underlying Ag, and thus the enrichment of Ag was more distinctive. Hence, AgCaLa was more likely positioned on the top and at the sides of a sample and chemical compounds were present on the top as well as at the side of the as-built samples.

Cross-sections verified the enrichment of AgCaLa on the top surface (Figure 3), but the AgCaLa layer was inhomogeneous as the layer was partly thin and elsewhere the AgCaLa was extended deep into the material. These deep incisive AgCaLa phases were positioned at the flanks of single melt pools, respectively, between adjacent melt tracks, as demonstrated by samples built without any rotation of the direction of scanning (Figure 3c,d). The deposition at the flanks of the melt pools was more distinctive for samples with a low density. Thus, AgCaLa agglomerated in the upper part of the samples, and the AgCaLa phases in the bulk material originated from these agglomerations. For scanning without any rotation of the direction of scanning, the connection between neighboring melt tracks was nearly lost and the material was not suited for the intended application (Figure 3d). This effect was not observed if a rotation was conducted (Figure 3b).

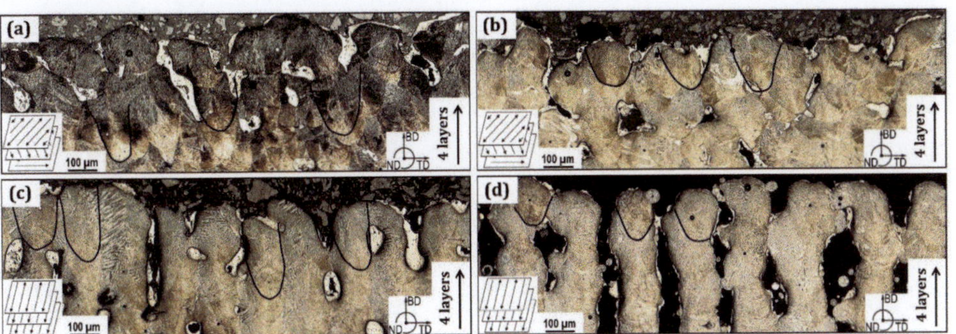

Figure 3. LM of etched cross-sections of top surfaces with enrichment of AgCaLa: (**a**) HD, (**b**) LD, (**c**) HD-0, AgCaLa phases between adjacent melt pools, narrow shape of melt pools, (**d**) LD-0.

Compared to samples with a high density (Figure 3a), the enrichment of AgCaLa on the upper surface, and in particular the deep incisive phases, was less distinctive for samples manufactured with the LD strategy. Thus, the conducting of the LD strategy and acceptance of a certain porosity reduces the enrichment of AgCaLa in the upper part and consequently seems to facilitate the incorporation of AgCaLa.

The observed AgCaLa phases did not have the bright appearance which is typical of Ag. Consequently, these phases did not match the observed slag on the top surface. Thus, the slag was a thin layer on the AgCaLa phases, but not a voluminous layer. As the EDS measurements reflected the chemical composition near the surface, including the slag, an even more distinctive enrichment of AgCaLa under the slag can be assumed than indicated by the EDS measurement in Figure 2a.

Aside from the observation of the morphology of the AgCaLa phases, the appearance of the melt pools was identifiable. The melt pools indicated in Figure 3a,c for the HD strategy had a relatively narrow and deep shape. In contrast, the melt pools resulting from the LD strategy were less deep and narrow (Figure 3b,d). Furthermore, the LBM-typical columnar dendritic structure of solidification was identifiable [83].

Investigations of the morphology of the AgCaLa phases within the bulk material demonstrated that the observed deep incisive AgCaLa phases at the side of melt pools continued into the bulk material, as the cross-sections of the bulk material without any rotation show aligned AgCaLa phases (Figure 4a). These AgCaLa phases were positioned between adjacent melt pools confirmed by the distribution of AgCaLa at the edge of the samples (Figure 4m). Additionally, the slightly visible grain structure confirmed the position between the melt pools in the bulk material as the grains grew from the side to the center of the melt pools, as indicated by arrows in Figure 4i [63,84]. An alignment along the melt tracks was also indicated by a section perpendicular to the building direction (Figure 4e). The positioning between adjacent tracks was more distinctive for material processed with the LD-0 strategy (Figure 4c,g,k,o). These samples had an almost continuous Ag layer at the edge of each melt track. However, these distinctive AgCaLa layers did not occur for the HD-0 strategy. For the HD-0 strategy, the AgCaLa was divided into separated phases (Figure 4a,e,i,m). This effect was more pronounced for the strategy HD with rotation of the scanning direction, as this material was characterized by homogenous AgCaLa phases with a compact shape (Figure 4b,f,j,n).

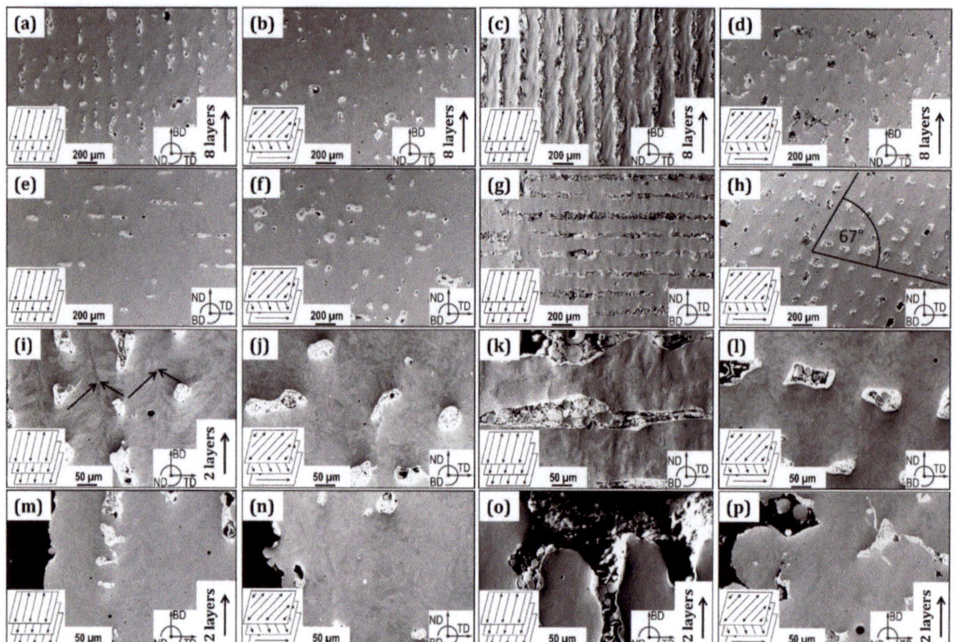

Figure 4. SEM-SE: morphology of AgCaLa phases in a FeMn matrix: (**a,e,i,m**) HD-0, AgCaLa phases positioned between adjacent melt pools; (**b,f,j,n**) HD, homogeneously distributed AgCaLa phases; (**c,g,k,o**) LD-0, distinctive deposition of AgCaLa phases between adjacent melt tracks with a weak connection between single tracks; (**d,h,l,p**) LD, AgCaLa phases aligned along the melt tracks, separated by the deep center of the melt pool of the following layer.

The intensive deposition of AgCaLa at the sides of the melt tracks matched the increased amount of Ag detected at the sides of as-built samples (Figures 2b and 4m,n,p). Distinctive AgCaLa layers in the bulk material were only observed for the LD-0 strategy and not for samples produced with the LD strategy with rotation of the scanning direction (Figure 4d,h,l,p). The AgCaLa phases of LD samples revealed regular structures and the phases were separated by continuous lines of FeMn. As the AgCaLa was preferred deposited at the side of melt tracks and these structures had an angle of 67°, corresponding to the rotation of the scanning direction, these phases were divided into smaller fractions by the deep center of the melt tracks of the following layers (Figure 4h,l).

Apart from the morphology, a porosity was specifically observed in samples produced with the LD and LD-0 strategies, whereby even samples with HD-0 and HD strategies exhibited a certain porosity. The pores were preferably located near the AgCaLa phases or within the AgCaLa phases. Aside from the porosity, further inclusions were present in most of the AgCaLa phases. These inclusions were in samples with high as well as low density.

Three-dimensional investigations of the morphology of AgCaLa in the bulk material confirmed a homogenous distribution for HD and LD samples within the whole volume (Figure 5d,e). For samples built with the LD strategy, the porosity was significantly pronounced. The acceptance of this certain porosity enabled the adjustment of smaller AgCaLa phases (Figure 5e). Thus, taking into account the previous results, well-dispersed AgCaLa phases could be obtained via LBM and the adjustment of regular AgCaLa phases was possible.

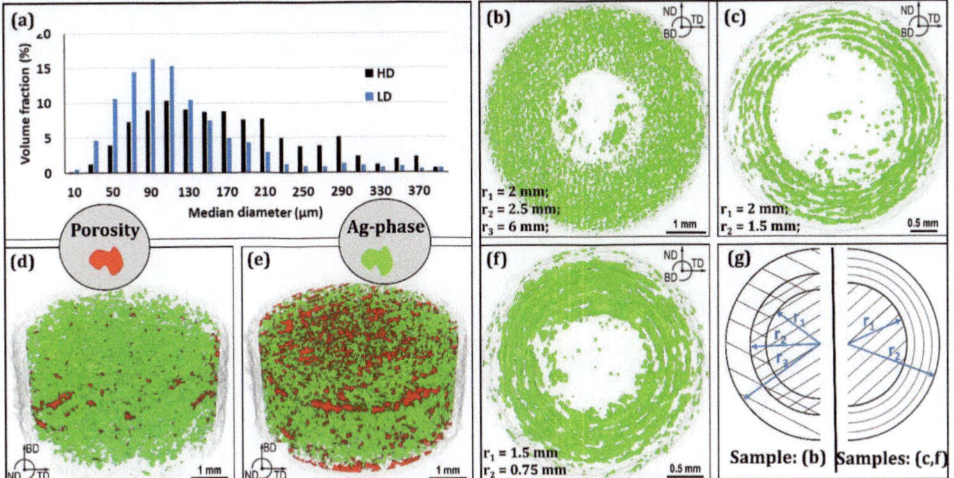

Figure 5. 3D-distribution of AgCaLa phases analyzed via μ-CT: (**a**) size distribution of AgCaLa phases; (**d**) HD, homogenous distribution of AgCaLa phases; (**e**) LD, homogenous distribution of AgCaLa phases and pores; (**b,c,f**) samples with a graduated distribution of AgCaLa phases by variation of LBM parameters; (**g**) schematic illustration of the parameter variation used for samples (**b,c,f**).

However, for both materials, some AgCaLa phases were significantly larger than the initial powder particles, indicating agglomeration of AgCaLa. These agglomerations confirmed the formation of the AgCaLa phases from the enrichment of AgCaLa in the upper part of the LBM manufactured parts (Figure 5a). Thus, the AgCaLa phases in the bulk did not originate from single powder particles.

As Figure 1g depicts, it was possible to produce samples without AgCaLa in the bulk material with LSC parameters. If these LSC parameters were applied for the center parts, an AgCaLa-free center could be obtained. Figure 5g demonstrates the applied hatchings schematically. HD and LSC parameters with an overlap resulted in an almost AgCaLa-free center and homogenously dispersed AgCaLa phases in the outer area (Figure 5b). Similar results were obtained when the HD parameters were used for several circular counter-paths in combination with LSC parameters for the center (Figure 5c,f). For the counter-paths, the stripe-like hatching is displaced by several circles (Figure 5g). This strategy resulted in AgCaLa phases aligned along the circular melt tracks comparable to the HD-0 structure (Figure 4e). In summary, it was possible to obtain well-dispersed AgCaLa phases in the outer area while the center was almost free of AgCaLa with both hatching strategies. Thus, the targeted adjustment of the homogenous distribution of AgCaLa phases was possible. Furthermore, an adaption of the morphology and size distribution was possible, as the different particle sizes of LD and HD parameter settings prove. As reported by Krüger et al. [35], further parameters such as layer thickness can be applied for further adaptions if the results of degradation tests, investigations of biocompatibility, etc., make adaptions necessary.

The corresponding investigation of the grain structure and distribution of the chemical elements point out the embedding of the AgCaLa phases in an austenitic matrix, which is typical for this material processed via LBM (Figure 6a). The columnar dendritic structure of solidification was detectable in the EBSD data and the associated segregations of Mn were slightly visible (Figure 6b) [35,83]. Segregations of Mn occurred on a larger scale, e.g., along the melt pool boundaries, as identified in Figure 6a [35,85].

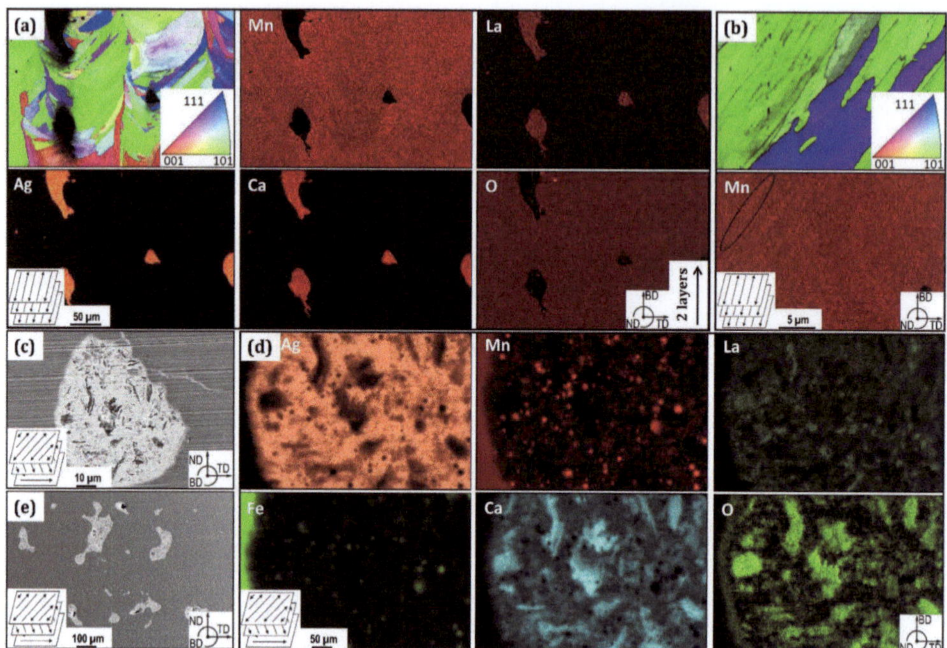

Figure 6. Microstructure of the FeMn matrix and AgCaLa phases: (**a**) SEM-EBSD with corresponding SEM-EDS: HD, austenitic matrix with AgCaLa phases; (**b**) SEM-EBSD with corresponding SEM-EDS: HD, columnar dendritic structure with segregations; (**c,e**) SEM-SE: HD, water-free prepared AgCaLa phases in the FeMn matrix; (**d**) SEM-EDS: HD, water-free prepared inhomogeneous distribution of alloying elements and O in an AgCaLa phase.

As expected and intended, the alloying elements Ca and La were detected within the Ag-rich areas, whereas Mn was detected within the matrix material (Figure 6a). However, it must be considered that the preparation might influence the chemical composition due to the reactivity of the alloying elements. Particularly, the final vibration polishing for 12 h might be crucial. Thus, for the investigation of the structure of the AgCaLa phases, all samples were solely ground to a grid size of 2500 using ethanol as a cooling lubricant. These samples showed less porosity and various inclusions in the AgCaLa phases instead of the porosity observed after the water-based preparation route (Figure 6c,e). The inclusions were enriched with Ca, O, and partially La (Figure 6d) and might be oxides of Ca. As a similar chemical composition was detected for the slag on the as-built top surface, these phases might be slag incorporated into the AgCaLa phases during LBM. The dissolution of these inclusions during preparation with water was conclusive since Ca forms hydroxides which are dissolvable in water but insoluble in alcohol [86]. Thus, the observed porosity (Figure 4) was due to the dissolving of Ca- and O-rich inclusions, and the real structure in the bulk material was better preserved in water-free prepared samples (Figure 6d). Additionally, to the Ca- and O-rich areas, small inclusions of FeMn were dispersed in the AgCaLa phases. However, between these inclusions, wide Ag-rich areas containing Ca and La were detected, as intended for the AgCaLa phase.

The preparation routine also influenced the O distribution significantly as O was mainly detected in the AgCaLa phases after the water-free preparation and in the area of bulk material after the water-based preparation. Thus, during a water-based preparation, oxides were formed in the area of the FeMn while the O-rich regions of AgCaLa were dissolved and the O distribution shown in Figure 6a was less representative than in Figure 6d.

For the AgCaLa phases of water-free prepared samples which were ground to 2500er grid size and subsequently polished with 3 μm water-free suspension, a composition of 75.8 wt.% Ag, 10.6 wt.% Ca, 5.4 wt.% La, 5.2 wt.% Mn, 3.9 wt.% O, and 0.7 wt.% Fe was determined via WDX (average of five particles with 121 measuring points each, outliers excluded). Thus, the alloying elements remained in the AgCaLa and the intended chemical composition of the AgCaLa was almost reached. However, the Ca content was less than in the powder feedstock and less than intended. Additionally, the detected O confirmed the formation of O-containing compounds, and some Mn was taken up from the surrounding matrix by the AgCaLa. The uptake of Mn was more pronounced for the processing of FeMn with pure Ag, as described by Krüger et al. [35], resulting in the chemical composition of 88.3 wt.% Ag, 11.7 wt.% Mn, 0.1 wt.% O, and 0.6 wt.% Fe (WDX).

The lack of any gradual transition of the chemical composition or enrichments of elements at the boundary between AgCaLa and FeMn indicates the lack of any connection or transition zone between both phases. Thus, the AgCaLa phases were present in the matrix without any connection to the matrix.

The fraction area resulting from the quasi-static tensile test confirmed the missing interconnection between the AgCaLa phases and the FeMn matrix since a detachment of AgCaLa, as well as cavities from which AgCaLa was detached, were observed (Figure 7b,c,d). Besides, brittle fractured AgCaLa phases were also present. These fractures occurred probably due to mechanical clamping with undercuts, albeit, even for fractured AgCaLa, a separation from the surrounding material was detected (Figure 7c). As expected and in contrast to the AgCaLa, the FeMn matrix was characterized by a ductile fracture behavior (Figure 7e) [19]. On a larger scale, the fracture was cleaved comparable to the typical behavior of brittle fracture behavior (Figure 7b,d). This structure was conclusive since the AgCaLa phases that did not carry a significant load were the weakest points and the fracture lines proceeded between them.

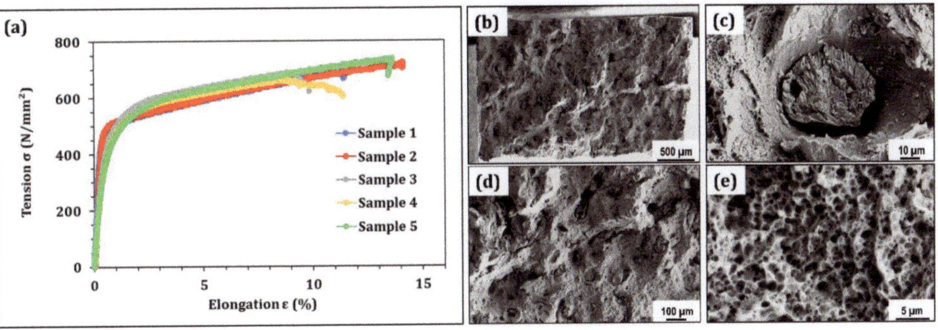

Figure 7. Results of tensile tests: (**a**) Tension elongation diagram of five samples, each represented by a different color; (**b–e**) SEM-SE: fracture area; (**b,d**) cleaved fracture, reduced load-bearing cross-section due to AgCaLa phases; (**c**) brittle fraction of AgCaLa phase; (**e**) ductile fraction of FeMn matrix.

The observed fracture surface is conclusive with the mechanical behavior, since after exceeding the yield strength, the material was 10% plastically deformed in agreement with the ductile structure of the matrix, while the failure occurred spontaneously without significant necking (Figure 7b). This failure behavior, which is more likely for brittle materials, was indicated by the lack of regression of tension before failure. Thus, the

FeMn enabled a ductile material behavior, but the notch effect of the AgCaLa phases was comparable to the influence of pores by suppressing necking [19,26,87].

The average ultimate tensile strength was 701 MPa (standard deviation: 29 MPa) and the elongation to fracture was 12.1% (standard deviation: 29 MPa). For some samples, the transition between elastic and plastic elongation was flattened and no distinct yield strength was distinguishable. Thus, a reliable determination of the yield strength was not possible, since the behavior of the samples significantly differed. For the samples with a distinct transition, a yield strength of 475 MPa can be estimated.

The following section aims to understand the influencing parameters regarding the formation and the resulting structure of the AgCaLa inclusions to enable further targeted adaption to create biodegradable Fe-based material. Furthermore, the suitability of the presented material for the intended application is discussed.

3.1. Formation of Insoluble AgCaLa Phases

The results presented demonstrate that a mixture of iron-based matrix material in combination with a degradable Ag alloy can be processed via LBM. A good and adaptable distribution of AgCaLa within the bulk material can be obtained and the chemical composition remains, although some minor changes due to loss and uptake of alloying elements and O occur. This is an outstanding result, as previous investigations revealed intensive interaction and diffusion of alloying elements between insoluble melts of FeMn and pure Ag during LBM [35].

However, the addition of 12 wt.% Ca and 5 wt.% La to Ag significantly influences the LBM process, although these alloying elements result in only 0.85 wt.% of the mixed material. In contrast to the processing of pure Ag with FeMn, where a transition welding mode is identified, a keyhole mode was determined for the present results due to the deep and narrow melt pools [35,62,63,80,81]. Here, the energy introduced with every single track was reduced for HD parameters compared to the parameters for the processing of FeMn, as well as FeMn with pure Ag. The hatch distance and scanning speed were increased while the laser power was decreased [35]. Thus, the different welding mode was not caused by increased energy introduction but by the alloying elements Ca and La. The absorption rate of AgCaLa was not the reason, as the absorption was lower for the processed material with AgCaLa. However, the formation of a keyhole due to vaporization increases the energy absorption. Consequently, the vaporization behavior of the material is decisive [65,66,88]. Ca has a significantly higher vapor pressure than Mn, Ag, Fe, and La [89]. Accordingly, the increasing vaporization and formation of keyholes due to the addition of even small amounts of Ca are conclusive. Additionally, the vapor pressure depends linearly on the mole fraction, so Ca has a decisive influence as the vapor pressure of Ca is at 1000 °C 10,000 times higher than for Mn [89,90]. In addition, the enrichment of AgCaLa in the upper part of the LBM manufactured material increases the influence of Ca. Accordingly, for the LD parameters, less enrichment of AgCaLa on the top was observed and the melt pools were not as deep and narrow.

As the keyhole mode increases the amount of energy available for melting, the melt pool size increases and it is to be expected that a reduced energy density is sufficient for the processing of FeMn with AgCaLa [64,65,67,68,88,91]. Further energy might be released by chemical reactions, indicated by the observed compounds on the top and within the AgCaLa phases [57]. Apart from the changed energy feeding, the heat dissipation might be affected due to differences in heat conductivity, radiation, etc. [53,59,92]. Nevertheless, the main influencing parameter is the high vapor pressure of Ca as it evokes the keyhole mode. Due to this, further minor differences in comparison to the processing of FeMn with pure Ag cannot be identified.

However, apart from the changing welding mode, similar formation of the AgCaLa phases compared to pure Ag phases as described by Krüger et al. [35] can be assumed. Hence, the model of the formation of AgCaLa phases presented in Figure 8 is based on conclusions from the results discussed here and the adaption of information from the literature, e.g., regarding melt flow and shape of the melt pool. In the first part of the melt pool, the recoil pressure is decisive [53,64,66,80]. The AgCaLa originating from the powder feedstock as well as the enrichment of AgCaLa on the upper surface were melted and emerged into the FeMn (Figure 8b). The emerging is conclusive, as the recoil pressure causes strong melt flow around the keyhole and affects both components similarly [64,71,93]. Additionally, the enrichment of AgCaLa on the top surface in the interaction zone with the laser is implausible, as this would reduce the energy adsorption, since Ag has a high reflectivity. Moreover, the observed diffusion of alloying elements between the liquid components supports the theory of emerging, as the emerging simplifies diffusion due to an increased contact interface.

The surface tension is decisive in the backward part of the melt pool and causes, e.g., Marangoni flow [53,59,80,94]. The results presented reveal that no transition between AgCaLa and FeMn exists. Thus, a phase border with a certain surface tension exists between both melts. Consequently, the AgCaLa coagulates when the floating phases come in contact, as the coagulation reduces surface energy. The position of AgCaLa on the top can be explained by the lower surface tension of AgCaLa compared to FeMn. As less surface energy is necessary to create a top surface of AgCaLa than for a top surface of FeMn, additional energy would be necessary for the incorporation of AgCaLa from the top surface to generate a free top surface of FeMn as well as a new surface between AgCaLa and FeMn. Accordingly, once AgCaLa reaches the surface it is trapped there. The lower surface tension of AgCaLa compared to FeMn is consistent, as the surface tension of Ag can be expected to be lower than that of FeMn [35,95,96]. Additionally, the surface tension of La and Ca is lower than for Ag. Since typically the surface tension of alloys ranges between the surface tension of the single elements, the surface tension of AgCaLa can be assumed to be lower than that of pure Ag [97–99]. Moreover, the position of AgCaLa on the top enables intense contact of AgCaLa with the atmosphere and uptake of O, which results in further reduction of surface tension [95].

AgCaLa reaches the top surface due to the random flow of the floating AgCaLa, as the melt flow is strong enough to keep the AgCaLa floating even against gravity [100,101]. A further effect supports the rising of AgCaLa particles: as the surface tension of the surrounding FeMn decreases with increasing temperature, the surface tension of the FeMn melt is the lowest in the upper and hottest part of the melt pool. Consequently, the FeMn flows under the floating AgCaLa towards higher surface tension, and the AgCaLa is pushed upwards [35,102,103]. Thus, the position of AgCaLa on the top is conclusive.

Another effect results in the position of the AgCaLa between neighboring melt tracks: The temperature depending on the surface tension induces an outward-directed melt flow vortex of the FeMn melt under AgCaLa [71,104–106]. This outward and downward directed flow drags the AgCaLa melt down (Figure 8a). The AgCaLa at the flanks of the melt pools is more easily detectable for the process strategies without any rotation of the scanning direction. The positioning at the sides of the melt pool is significantly more pronounced for AgCaLa than for Ag [35]. This might be due to different melting modes or could be influenced by the different material properties of AgCaLa. The bigger difference in surface tension between AgCaLa and FeMn than between FeMn and Ag could result in a stronger manifestation of the surface tension-driven mechanisms observed for pure Ag. Furthermore, the AgCaLa enables the use of parameters with increased hatch distance compared to a combination of FeMn with Ag [35]. An increased hatch distance enables easier incorporation of the AgCaLa, as the LD parameters prove. With such an LD strategy, a regular structure of AgCaLa phases arranged along the melt track and divided into insular phases by the deep center of the following melt track can be obtained. These unique structures confirm the effects responsible for the formation of the AgCaLa phases. Thus, the presented results match the previous results for the processing of FeMn with pure Ag by Krüger et al. [35], and the more significant morphology of the AgCaLa phases confirms the stated hypotheses regarding the Ag phase formation.

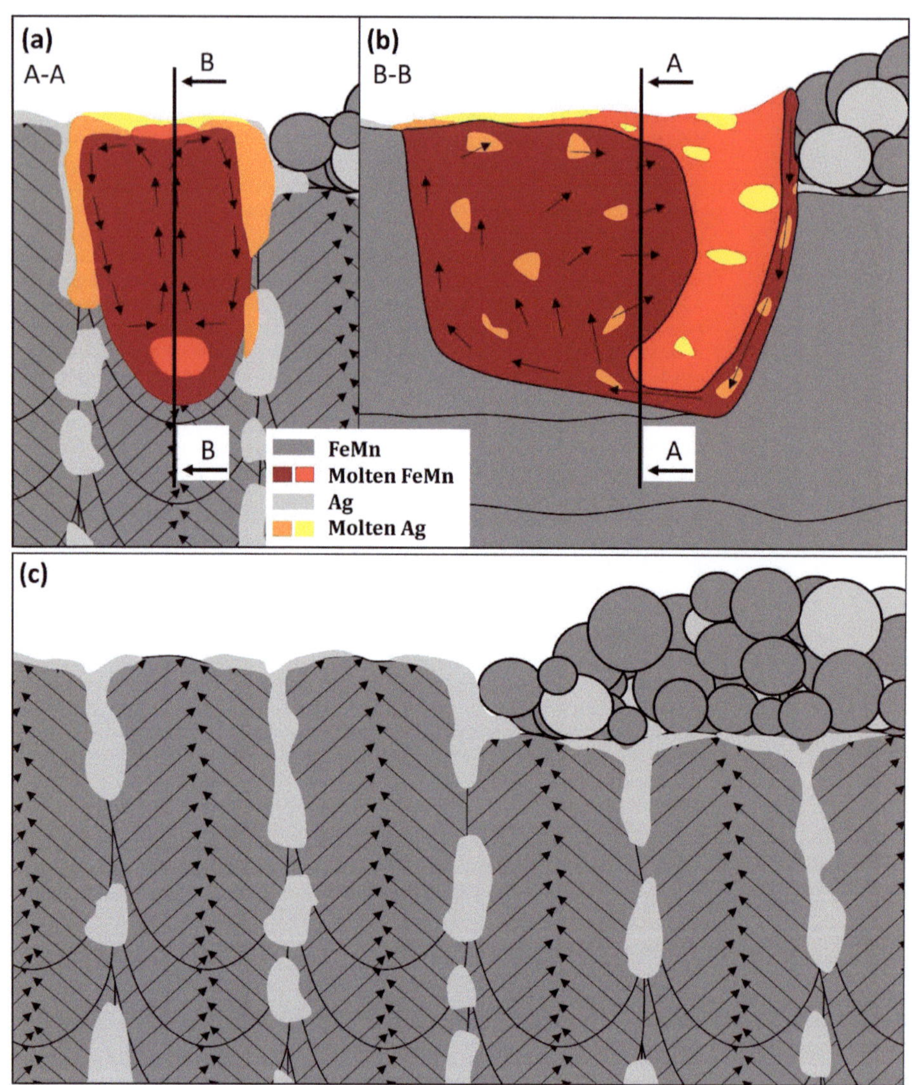

Figure 8. Formation of AgCaLa phases throughout the LBM process, schematic model; melt pool shape and melt flow adapted from [60,64,68,71,93,107]: (**a**) outward directed vortex of FeMn transporting AgCaLa to the flanks of the melt pool; (**b**) emerging of AgCaLa due to keyhole and accumulation on the top in the backward part of the melt pool due to surface tension; (**c**) distribution of AgCaLa in the top of LBM-manufactured material, with remaining AgCaLa phases at the lower end of the melt pools.

As the AgCaLa is transported to the top of the melt pool, the AgCaLa in the upper part of the melt pool is remelted and transported upwards again with the generation of each new layer. This mechanism leads to an increased amount of AgCaLa in the melt pool. Only AgCaLa phases at the lower end of the melt pool remain in the bulk material during the generation of the next layer. After a few layers, an equilibrium in the addition of AgCaLa from the powder feedstock and the remaining of AgCaLa in the bulk material is achieved. Thus, the AgCaLa phases in the bulk material are remaining parts of the agglomeration of AgCaLa on the top and do not originate from initial powder particles. This agglomeration, mainly caused by surface tension-driven effects, is conclusive with the determined size of the AgCaLa phases (Figure 5a) since some phases are larger than initial powder particles.

3.2. Spattering Causing Graded Distribution of AgCaLa

The model presented cannot explain why in some samples no AgCaLa phases were formed. According to the model, even if the accumulation on the top were more pronounced, after several layers too much AgCaLa would be accumulated and remain in the bulk material. However, less AgCaLa was observed on the top of samples manufactured with LSC parameters. Therefore, the AgCaLa must be removed by other effects. One explanation might be the excessive formation of AgCaLa spatters. As specified in the introduction, the spattering depends, e.g., on the vaporization behavior, viscosity, and surface tension of a material [57,60,72]. Hence, the intensive spattering of AgCaLa due to the high vapor pressure of Ca is conclusive, as the evaporation causes recoil and accelerates the material. Additionally, the reduced surface tension of AgCaLa might simplify the shearing of AgCaLa. However, the significant influence of the material properties on the spattering is observed by Gunenthiram et al. [60].

Apart from ejection from the melt pool, an increased ejection of AgCaLa from the powder bed is possible as evaporation can create a blow-off impulse [57,60,62,70,82]. Since Ca evaporates excessively, an increased blow-off impulse concerning the AgCaLa particles is conclusive. Thus, the preferred ejection of AgCaLa is conclusive but occurs only for parameter settings which lead to a high energy density. A reduced scanning speed and increased laser power (LSC-1 and LSC-2) increase the energy in the melt pool, resulting in increased spattering [57,60,70]. Nevertheless, regarding the scanning speed opposite effects are reported [57,70]. However, in the present study, the LSC parameters led to increased spattering and, as AgCaLa was more likely to be removed from the melt pool, material that did not contain any AgCaLa could be generated. Additionally, some of the effects resulting in increased spattering of AgCaLa might only occur at a higher energy level.

However, the presented results prove the pronounced sensitivity to the process parameters, which can be utilized to create material with graded properties. Thus, it is possible to create components with AgCaLa phases only in the defined areas. To reduce the degradation rate at the beginning, fewer AgCaLa phases in the edged area are beneficial. If instead of the degradable FeMn a not degradable Fe-based matrix material like 316L is used, it would be possible to manufacture implants that release Ag. As Ag acts antibacterially, those implants could prevent implant-related infections [40,45,108,109]. In such material, the mechanical load capability would be less affected due to the lack of AgCaLa in the center. Thus, the results achieved in this study can be applied to other applications of immiscible elements in multi-material systems.

3.3. Generation of Slag and Chemical Composition of AgCaLa

Apart from the possibility to generate a suitable distribution of AgCaLa phases, the chemical composition of the AgCaLa phases has to be discussed critically. The composition is changed by the loss of Ca and the intake of Mn. However, knowing this, a potential loss of Ca can be balanced by an increased content in the feedstock, whereas the intake of Mn is less distinctive than for pure Ag [35]. Thus, the addition of Ca and La reduces the solubility of Mn in Ag. The reduced intake of Mn proves the possibility to prevent the uptake of undesirable elements by the addition of acceptable elements. This option is considerably important for degradable Ag phases within a not degradable matrix, as the intake of, e.g., nickel (Ni) is critical for the human organism.

Apart from the deviation of the overall chemical composition, the inclusion of chemical compounds generated during the LBM process might be a crucial point. The occurrence of such reactions despite the low oxygen content during LBM is a common challenge, specifically for the processing of alloys with reactive elements [73–75,79]. When this slag is incorporated into the material it can weaken the mechanical properties [76,78]. In the present case, the investigated material contained slag only in combination with AgCaLa phases which did not contribute to the load capacity. The incorporation of the slag into AgCaLa is conclusive due to the lower surface tension and melting temperature of AgCaLa. Since the oxides of Mn decompose and melt at high temperatures, incorporation to FeMn with high melting temperature is less probable [46,110,111]. Additionally, the slag is located on the top of AgCaLa, and AgCaLa lies on FeMn; thus, the slag is in contact with AgCaLa and not with FeMn. The low density contributes to the fact that AgCaLa is on the top [100,104]. Anyway, in comparison to the distinctive slag on the top, only a little amount of slag is detected in the bulk material. This might be due to repeated accumulation of slag on the top accompanied by the ejection of oxides with the spatters, the simultaneous oxidation and de-oxidation, as well as vaporization of oxides in the laser spot, which might contribute to the reduced incorporation of slag into the bulk material [69,73,76,112]. Aside from the incorporation of slag, porosity is observed together with the AgCaLa in the bulk material, even for the HD parameter settings. This porosity should be noncritical, as improved osteoconductivity and degradation behavior are reported for targeted porosity [16,113]. Thus, the porosity caused by the AgCaLa might be beneficial.

However, the AgCaLa phases are suited to the intended application despite the deviation in the chemical composition, as Ag-rich areas are present. These regions presumably have a higher electrochemical potential than the surrounding matrix and should therefore cause enhanced anodic dissolution of the FeMn. The inclusions of slag might be dissolved directly after contact with the aqueous environment, as the samples prepared with water reveal. If these inclusions remain, they could enhance the dissolution of the AgCaLa phases due to differences in the electrochemical potential. The intake of Mn might also contribute to the degradation of the AgCaLa phases after the dissolution of the matrix material. Otherwise, Mn reduces the electrochemical potential of the AgCaLa phases. Whether this reduction is critical depends on the extent of the reduction. Therefore, investigations of the degradation behavior are mandatory to determine the influence of the AgCaLa phases, as presented in this study, and enable further defined adaption, e.g., by adjusting the morphology via process conditions and the chemical composition via the feedstock material.

3.4. Mechanical Properties Affected by AgCaLa

Apart from the degradation behavior, the material needs to fulfill the mechanical requirements. The results of the quasi-static tensile test are promising as the yield strength of approximately 475 MPa is only 5% lower than that of additively processed FeMn, and FeMn modified with 5 wt.% pure Ag. The modification with pure Ag impacts the yield strength just slightly. The elongation at break (identical sample shape) of FeMn is between 25% and 45%, and 16% and 24% for FeMn with pure Ag. Thus, the elongation at break is significantly reduced to 12.1% for the modification with AgCaLa [35]. As any deformation of an implant is not acceptable, the elongation at break is not considered to be crucial. However, a certain ductile material behavior is recommended since for a coincidental overload the implant would not break directly. For the ultimate tensile strength, the identical tendency as for yield strength and elongation at break is observed. The ultimate tensile strength is about 1025 MPa for FeMn, 850 MPa for FeMn modified with pure Ag, and 701 MPa for FeMn modified with AgCaLa. Thus, the modification with pure Ag affects the mechanical properties less than the modification with AgCaLa [35]. The volume fraction of AgCaLa is higher than for Ag due to the lower density of AgCaLa and thus the impact of AgCaLa on the load carrying section is increased. Furthermore, the reduction observed for FeMn with AgCaLa might be due to the porosity observed in combination with the AgCaLa phases, as the load-bearing cross-section is further reduced. Porosity could be related to chemical reactions and slag incorporation since porosity is observed in combination with slag in single-component materials [78]. This increased porosity might also be an explanation for the flattened transition between elastic and plastic deformation. However, the porosity might be beneficial to adapt the stiffness to the stiffness of bones and reduce the stress shielding effect [51,52].

In summary, the quasi-static mechanical properties are promising apart from a few restrictions which can be improved by process adaption to reduce the generation of slag. As fatigue behavior is a decisive factor for medical implants, it has to be investigated additionally. It can be assumed that the notch effect of the AgCaLa phases in combination with pores affects the fatigue behavior more strongly than the quasi-static properties. However, the AgCaLa phases exhibit a globular shape. Together with the ductile behavior of the matrix material, a reduction of tension peaks can lead to an acceptable fatigue behavior.

4. Conclusions

The present study reports on the processing of FeMn with a degradable Ag alloy via LBM for the adaption of the degradation rate. The effects influencing the morphology and chemical composition of the Ag phases are identified to enable further targeted adaption for the application as implant material. Additionally, the suitability of the material as degradable implant material is discussed.

LBM enables the processing of FeMn with degradable AgCaLa alloy. Suited, adaptable morphology and chemical composition of the AgCaLa phases are obtained.

AgCaLa is enriched on the top and between adjacent melt tracks due to surface tension-driven effects.

AgCaLa in the upper part is remelted; only AgCaLa at the lower end of the melt pools remains in the bulk material.

Graded distributions with regular structures and regions without AgCaLa can be achieved as the morphology of AgCaLa phases is sensitive to LBM parameters.

The use of AgCaLa instead of pure Ag significantly influences the LBM process. The high vapor pressure of Ca causes keyhole welding mode.

Slag is generated due to the high reactivity of Ca, La, and Mn. The slag is incorporated into the AgCaLa phases. The Ag phases take up Mn from the matrix.

Suited areas with high Ag content in the AgCaLa phases should enable increased anodic dissolution of the FeMn matrix due to different electrochemical potentials.

Due to the addition of AgCaLa, the yield strength is reduced by only 5% compared to pure FeMn.

As the presented results demonstrate, the chemical composition of an Ag alloy processed via LBM with an iron-based matrix influences the process characteristics and the material properties in many aspects. Thus, all relevant aspects, such as changes in the welding mode, the loss and intake of alloying elements by diffusion, varying saturation concentration, vaporization of alloying elements, or generation of slag have to be considered to generate a biocompatible material with a suited morphology of Ag phases and degradation behavior. However, the material presented in this study reveals promising properties. Investigations of the degradation behavior and biocompatibility are required to enable further targeted adaptions to be made.

Funding: This research was funded by German Research Foundation (DFG) grant number SCHA1484/44-1.

Acknowledgments: The author gratefully thanks the German Research Foundation (DFG) for financial support. This work results from the project SCHA1484/44-1. Additionally, the author thanks Mirko Schaper and Kay-Peter Hoyer for enabling this work and supervision. Furthermore, the author would like to thank Anatolii Andreiev, Sabrina Beumer, and Lea Kaspersmeier for promoting the production and examination of the tested material.

Conflicts of Interest: The author declares to have no conflict of interest.

References

1. Zivic, F.; Affatato, S.; Trajanovic, M.; Schnabelrauch, M.; Grujovic, N.; Choy, K.L. *Choy Biomaterials in Clinical Practice*; Springer: Cham, Switzerland, 2018.
2. Chen, Q.; Thouas, G.A. Metallic implant biomaterials. *Mater. Sci. Eng. R Rep.* **2015**, *87*, 1–57. [CrossRef]
3. Carluccio, D.; Demir, A.G.; Bermingham, M.J.; Dargusch, M.S. Challenges and Opportunities in the Selective Laser Melting of Biodegradable Metals for Load-Bearing Bone Scaffold Applications. *Metall. Mater. Trans. A* **2020**, *51*, 3311–3334. [CrossRef]
4. Garcia, D.R.; Deckey, D.G.; Zega, A.; Mayfield, C.; Spake, C.S.L.; Emanuel, T.; Daniels, A.; Jarrell, J.; Glasser, J.; Born, C.T.; et al. Analysis of growth and biofilm formation of bacterial pathogens on frequently used spinal implant materials. *Spine Deform.* **2020**, *8*, 351–359. [CrossRef] [PubMed]
5. Hermawan, H. Updates on the research and development of absorbable metals for biomedical applications. *Prog. Biomater.* **2018**, *7*, 93–110. [CrossRef] [PubMed]
6. Veerachamy, S.; Yarlagadda, T.; Manivasagam, G.; Yarlagadda, P.K.D.V. Bacterial adherence and biofilm formation on medical implants: A review. *Proc. Inst. Mech. Eng. Part H J. Eng. Med.* **2014**, *228*, 1083–1099. [CrossRef] [PubMed]
7. Zhang, E.; Zhao, X.; Hu, J.; Wang, R.; Fu, S.; Qin, G. Antibacterial metals and alloys for potential biomedical implants. *Bioact. Mater.* **2021**, *6*, 2569–2612. [CrossRef] [PubMed]
8. Zheng, Y.F.; Gu, X.N.; Witte, F. Biodegradable metals. *Mater. Sci. Eng. R Rep.* **2014**, *77*, 1–34. [CrossRef]
9. Redlich, C.; Schauer, A.; Scheibler, J.; Poehle, G.; Barthel, P.; Maennel, A.; Adams, V.; Weissgaerber, T.; Linke, A.; Quadbeck, P. In Vitro Degradation Behavior and Biocompatibility of Bioresorbable Molybdenum. *Metals* **2021**, *11*, 761. [CrossRef]
10. Hermawan, H.; Dubé, D.; Mantovani, D. Development of Degradable Fe-35Mn Alloy for Biomedical Application. *Adv. Mater. Res.* **2007**, *15–17*, 107–112. [CrossRef]
11. Abbaspour, N.; Hurrell, R.; Kelishadi, R. Review on Iron and Its Importance for Human Health. *J. Res. Med. Sci.* **2014**, *19*, 164–174.
12. Dargusch, M.S.; Venezuela, J.; Dehghan-Manshadi, A.; Johnston, S.; Yang, N.; Mardon, K.; Lau, C.; Allavena, R. In Vivo Evaluation of Bioabsorbable Fe-35Mn-1Ag: First Reports on In Vivo Hydrogen Gas Evolution in Fe-Based Implants. *Adv. Healthc. Mater.* **2020**, *10*, 2000667. [CrossRef]
13. Lin, W.; Qin, L.; Qi, H.; Zhang, D.; Zhang, G.; Gao, R.; Qiu, H.; Xia, Y.; Cao, P.; Wang, X.; et al. Long-term in vivo corrosion behavior, biocompatibility and bioresorption mechanism of a bioresorbable nitrided iron scaffold. *Acta Biomater.* **2017**, *54*, 454–468. [CrossRef]
14. Peuster, M.; Hesse, C.; Schloo, T.; Fink, C.; Beerbaum, P.; von Schnakenburg, C. Long-term biocompatibility of a corrodible peripheral iron stent in the porcine descending aorta. *Biomaterials* **2006**, *27*, 4955–4962. [CrossRef]
15. Dargusch, M.S.; Dehghan-Manshadi, A.; Shahbazi, M.; Venezuela, J.; Tran, X.; Song, J.; Liu, N.; Xu, C.; Ye, Q.; Wen, C. Exploring the Role of Manganese on the Microstructure, Mechanical Properties, Biodegradability, and Biocompatibility of Porous Iron-Based Scaffolds. *ACS Biomater. Sci. Eng.* **2019**, *5*, 1686–1702. [CrossRef]
16. Mandal, S.; Viraj; Nandi, S.K.; Roy, M. Effects of multiscale porosity and pore interconnectivity on in vitro and in vivo degradation and biocompatibility of Fe–Mn–Cu scaffolds. *J. Mater. Chem. B* **2021**, *9*, 4340–4354. [CrossRef]
17. Tonna, C.; Wang, C.; Mei, D.; Lamaka, S.V.; Zheludkevich, M.L.; Buhagiar, J. Biodegradation behaviour of Fe-based alloys in Hanks' Balanced Salt Solutions: Part I. material characterisation and corrosion testing. *Bioact. Mater.* **2022**, *7*, 426–440. [CrossRef]
18. Shuai, C.; Yang, W.; Yang, Y.; Pan, H.; He, C.; Qi, F.; Xie, D.; Liang, H. Selective laser melted Fe-Mn bone scaffold: Microstructure, corrosion behavior and cell response. *Mater. Res. Express* **2019**, *7*, 015404. [CrossRef]
19. Bouaziz, O.; Allain, S.; Scott, C.P.; Cugy, P.; Barbier, D. High manganese austenitic twinning induced plasticity steels: A review of the microstructure properties relationships. *Curr. Opin. Solid State Mater. Sci.* **2011**, *15*, 141–168. [CrossRef]
20. Hermawan, H.; Purnama, A.; Dube, D.; Couet, J.; Mantovani, D. Fe–Mn alloys for metallic biodegradable stents: Degradation and cell viability studies. *Acta Biomater.* **2010**, *6*, 1852–1860. [CrossRef]
21. Schinhammer, M.; Gerber, I.; Hänzi, A.C.; Uggowitzer, P.J. On the cytocompatibility of biodegradable Fe-based alloys. *Mater. Sci. Eng. C* **2013**, *33*, 782–789. [CrossRef] [PubMed]
22. Drynda, A.; Hassel, T.; Bach, F.W.; Peuster, M. In vitro and in vivo corrosion properties of new iron–manganese alloys designed for cardiovascular applications. *J. Biomed. Mater. Res. Part B Appl. Biomater.* **2014**, *103*, 649–660. [CrossRef] [PubMed]
23. Kraus, T.; Moszner, F.; Fischerauer, S.; Fiedler, M.; Martinelli, E.; Eichler, J.; Witte, F.; Willbold, E.; Schinhammer, M.; Meischel, M.; et al. Biodegradable Fe-based alloys for use in osteosynthesis: Outcome of an in vivo study after 52 weeks. *Acta Biomater.* **2014**, *10*, 3346–3353. [CrossRef] [PubMed]
24. Paul, B.; Lode, A.; Placht, A.-M.; Voß, A.; Pilz, S.; Wolff, U.; Oswald, S.; Gebert, A.; Gelinsky, M.; Hufenbach, J. Cell–Material Interactions in Direct Contact Culture of Endothelial Cells on Biodegradable Iron-Based Stents Fabricated by Laser Powder Bed Fusion and Impact of Ion Release. *ACS Appl. Mater. Interfaces* **2021**, *14*, 439–451. [CrossRef] [PubMed]
25. Schinhammer, M.; Steiger, P.; Moszner, F.; Löffler, J.F.; Uggowitzer, P.J. Degradation performance of biodegradable FeMnC(Pd) alloys. *Mater. Sci. Eng. C* **2013**, *33*, 1882–1893. [CrossRef]
26. Niendorf, T.; Brenne, F.; Hoyer, P.; Schwarze, D.; Schaper, M.; Grothe, R.; Wiesener, M.; Grundmeier, G.; Maier, H.J. Processing of New Materials by Additive Manufacturing: Iron-Based Alloys Containing Silver for Biomedical Applications. *Met. Mater. Trans. A* **2015**, *46*, 2829–2833. [CrossRef]

27. Čapek, J.; Msallamová, Š.; Jablonská, E.; Lipov, J.; Vojtěch, D. A novel high-strength and highly corrosive biodegradable Fe-Pd alloy: Structural, mechanical and in vitro corrosion and cytotoxicity study. *Mater. Sci. Eng. C* **2017**, *79*, 550–562. [CrossRef]

28. Mandal, S.; Kishore, V.; Bose, M.; Nandi, S.K.; Roy, M. In vitro and in vivo degradability, biocompatibility and antimicrobial characteristics of Cu added iron-manganese alloy. *J. Mater. Sci. Technol.* **2021**, *84*, 159–172. [CrossRef]

29. Huang, T.; Cheng, J.; Bian, D.; Zheng, Y. Fe-Au and Fe-Ag composites as candidates for biodegradable stent materials. *J. Biomed. Mater. Res. Part B Appl. Biomater.* **2016**, *104*, 225–240. [CrossRef]

30. Mijnendonckx, K.; Leys, N.; Mahillon, J.; Silver, S.; Van Houdt, R. Antimicrobial silver: Uses, toxicity and potential for resistance. *BioMetals* **2013**, *26*, 609–621. [CrossRef]

31. Krüger, J.T.; Hoyer, K.-P.; Filor, V.; Pramanik, S.; Kietzmann, M.; Meißner, J.; Schaper, M. Novel AgCa and AgCaLa alloys for Fe-based bioresorbable implants with adapted degradation. *J. Alloys Compd.* **2021**, *871*, 159544. [CrossRef]

32. Wiesener, M.; Peters, K.; Taube, A.; Keller, A.; Hoyer, K.-P.; Niendorf, T.; Grundmeier, G. Corrosion properties of bioresorbable FeMn-Ag alloys prepared by selective laser melting. *Mater. Corros.* **2017**, *68*, 1028–1036. [CrossRef]

33. Bagha, P.S.; Khakbiz, M.; Sheibani, S.; Hermawan, H. Design and characterization of nano and bimodal structured biodegradable Fe-Mn-Ag alloy with accelerated corrosion rate. *J. Alloys Compd.* **2018**, *767*, 955–965. [CrossRef]

34. Liu, R.-Y.; He, R.-G.; Chen, Y.-X.; Guo, S.-F. Effect of Ag on the Microstructure, Mechanical and Bio-corrosion Properties of Fe–30Mn Alloy. *Acta Met. Sin.* **2019**, *32*, 1337–1345. [CrossRef]

35. Krüger, J.T.; Hoyer, K.-P.; Hengsbach, F.; Schaper, M. Formation of insoluble silver-phases in an iron-manganese matrix for bioresorbable implants using varying laser beam melting strategies. *J. Mater. Res. Technol.* **2022**, *19*, 2369–2387. [CrossRef]

36. Hall, D.J.; Pourzal, R.; Jacobs, J.J.; Urban, R.M. Metal wear particles in hematopoietic marrow of the axial skeleton in patients with prior revision for mechanical failure of a hip or knee arthroplasty. *J. Biomed. Mater. Res. Part B Appl. Biomater.* **2019**, *107*, 1930–1936. [CrossRef]

37. Heiden, M.; Walker, E.; Stanciu, L. Magnesium, Iron and Zinc Alloys, the Trifecta of Bioresorbable Orthopaedic and Vascular Implantation—A Review. *J. Biotechnol. Biomater.* **2015**, *5*, 1000178. [CrossRef]

38. Afifi, M.; Saddick, S.; Zinada, O.A.A. Toxicity of silver nanoparticles on the brain of *Oreochromis niloticus* and *Tilapia zillii*. *Saudi J. Biol. Sci.* **2016**, *23*, 754–760. [CrossRef]

39. Andreiev, A.; Hoyer, K.P.; Grydin, O.; Frolov, Y.; Schaper, M. Degradable silver-based alloys. *Mater. Werkst.* **2020**, *51*, 517–530. [CrossRef]

40. Krüger, J.T.; Hoyer, K.P.; Schaper, M. Bioresorbable AgCe and AgCeLa alloys for adapted Fe-based implants. *Mater. Lett.* **2022**, *306*, 130890. [CrossRef]

41. Hoyer, K.P.; Schaper, M. Alloy Design for Biomedical Applications in Additive Manufacturing. In *TMS 2019 148th Annual Meeting & Exhibition Supplemental Proceedings*; The Minerals, Metals & Materials Series; Springer: Cham, Switzerland, 2019. [CrossRef]

42. Zhao, J.Z.; Ahmed, T.; Jiang, H.X.; He, J.; Sun, Q. Solidification of Immiscible Alloys: A Review. *Acta Metall. Sin.* **2017**, *30*, 1–28. [CrossRef]

43. ASM. *ASM Alloy Phase Diagram Database*; ASM International: Materials Park, OH, USA, 1999.

44. Conti, M.C.; Mallia, B.; Sinagra, E.; Wismayer, P.S.; Buhagiar, J.; Vella, D. The effect of alloying elements on the properties of pressed and non-pressed biodegradable Fe–Mn–Ag powder metallurgy alloys. *Heliyon* **2019**, *5*, e02522. [CrossRef]

45. Sotoudehbagha, P.; Sheibanib, S.; Khakbiz, M.; Ebrahimi-Barough, S.; Hermawand, H. Novel antibacterial biodegradable Fe-Mn-Ag alloys produced by mechanical alloying. *Mater. Sci. Eng. C* **2018**, *88*, 88–94. [CrossRef]

46. Kraner, J.; Medved, J.; Godec, M.; Paulin, I. Thermodynamic Behavior of Fe-Mn and Fe-Mn-Ag Powder Mixtures during Selective Laser Melting. *Metals* **2021**, *11*, 234. [CrossRef]

47. Quan, J.; Lina, K.; Gua, D. Selective laser melting of silver submicron powder modified 316L stainless steel: Influence of silver addition on microstructures and performances. *Powder Technol.* **2020**, *364*, 478–483. [CrossRef]

48. Guo, X.; Zhang, C.; Tian, Q.; Yu, D. Liquid metals dealloying as a general approach for the selective extraction of metals and the fabrication of nanoporous metals: A review. *Mater. Today Commun.* **2021**, *26*, 102007. [CrossRef]

49. Mikhailov, G.; Makrovets, L. Thermodynamic analysis of steel deoxidation with calcium and aluminum. *Russ. Metall.* **2008**, *2008*, 727–729. [CrossRef]

50. Garrison, W.M., Jr.; Maloney, J.L. Lanthanum additions and the toughness of ultra-high strength steels and the determination of appropriate lanthanum additions. *Mater. Sci. Eng. A* **2005**, *403*, 299–310. [CrossRef]

51. Leordean, D.; Dudescu, C.; Marcu, T.; Berce, P.; Balc, N. Customized implants with pecific properties, made by selective laser melting. *Rapid Prototyp. J.* **2015**, *21*, 98–104. [CrossRef]

52. Rodriguez-Contreras, A.; Punset, M.; Calero, J.A.; Gil, F.J.; Ruperez, E.; Manero, J.M. Powder metallurgy with space holder for porous titanium implants: A review. *J. Mater. Sci. Technol.* **2021**, *76*, 129–149. [CrossRef]

53. Khairallah, S.A.; Anderson, A.T.; Rubenchik, A.; King, W.E. Laser powder-bed fusion additive manufacturing: Physics of complex melt flow and formation mechanisms of pores, spatter, and denudation zones. *Acta Mater.* **2016**, *108*, 36–45. [CrossRef]

54. Heeling, T.; Cloots, M.; Wegener, K. Melt Pool Simulation for the Evaluation of Process Parameters in Selective Laser Melting. *Addit. Manuf.* **2017**, *14*, 116–125. [CrossRef]

55. Papazoglou, E.L.; Karkalos, N.E.; Karmiris-Obratański, P.; Markopoulos, A.P. On the Modeling and Simulation of SLM and SLS for Metal and Polymer Powders: A Review. *Arch. Comput. Methods Eng.* **2022**, *29*, 941–973. [CrossRef]

56. Tan, J.H.K.; Sing, S.L.; Yeong, W.Y. Microstructure modelling for metallic additive manufacturing: A review. *Virtual Phys. Prototyp.* **2019**, *15*, 87–105. [CrossRef]
57. Wang, D.; Wu, S.; Fu, F.; Mai, S.; Yang, Y.; Liu, Y.; Song, C. Mechanisms and characteristics of spatter generation in SLM processing and its effect on the properties. *Mater. Des.* **2017**, *117*, 121–130. [CrossRef]
58. Zhou, X.; Li, K.; Zhang, D.; Liu, X.; Ma, J.; Liu, W.; Shen, Z. Textures formed in a CoCrMo alloy by selective laser melting. *J. Alloys Compd.* **2015**, *631*, 153–164. [CrossRef]
59. Khairallah, S.A.; Anderson, A. Mesoscopic simulation model of selective laser melting of stainless steel powder. *J. Mater. Process. Technol.* **2014**, *214*, 2627–2636. [CrossRef]
60. Gunenthiram, V.; Peyre, P.; Schneider, M.; Dal, M.; Coste, F.; Koutiri, I.; Fabbro, R. Experimental analysis of spatter generation and melt-pool behavior during the powder bed laser beam melting process. *J. Mater. Process. Technol.* **2017**, *251*, 376–386. [CrossRef]
61. Matthews, M.J.; Guss, G.; Khairallah, S.A.; Rubenchik, A.M.; Depond, P.J.; King, W.E. Denudation of metal powder layers in laser powder bed fusion processes. *Acta Mater.* **2016**, *114*, 33–42. [CrossRef]
62. Gunenthiram, V.; Peyre, P.; Schneider, M.; Dal, M.; Coste, F.; Fabbro, R. Analysis of laser–melt pool–powder bed interaction during the selective laser melting of a stainless steel. *J. Laser Appl.* **2017**, *29*, 22303. [CrossRef]
63. Andreau, O.; Koutiri, I.; Peyre, P.; Penot, J.D.; Saintier, N.; Pessard, E.; de Terris, T.; Dupuy, C.; Baudin, T. Texture control of 316L parts by modulation of the melt pool morphology in selective laser melting. *J. Mater. Process. Technol.* **2019**, *264*, 21–31. [CrossRef]
64. Shrestha, S.; Chou, Y.K. A Numerical ASME Study on the Keyhole Formation During Laser Powder Bed Fusion Process. *J. Manuf. Sci. Eng.* **2019**, *141*, 101002. [CrossRef]
65. Huang, L.; Hua, X.; Wu, D.; Fang, L.; Cai, Y.; Ye, Y. Effect of magnesium content on keyhole-induced porosity formation and distribution in aluminum alloys laser welding. *J. Manuf. Process.* **2018**, *33*, 43–53. [CrossRef]
66. Queva, A.; Guillemot, G.; Moriconi, C.; Metton, C.; Bellet, M. Numerical study of the impact of vaporisation on melt pool dynamics in Laser Powder Bed Fusion—Application to IN718 and Ti–6Al–4V. *Addit. Manuf.* **2020**, *35*, 101249. [CrossRef]
67. Miyagi, M.; Wang, H.; Yoshida, R.; Kawahito, Y.; Kawakami, H.; Shoubu, T. Effect of alloy element on weld pool dynamics in laser welding of aluminum alloys. *Sci. Rep.* **2018**, *8*, 12944. [CrossRef]
68. Rehman, A.U.; Mahmood, M.A.; Pitir, F.; Salamci, M.U.; Popescu, A.C.; Mihailescu, I.N. Keyhole Formation by Laser Drilling in Laser Powder Bed Fusion of Ti6Al4V Biomedical Alloy: Mesoscopic Computational Fluid Dynamics Simulation versus Mathematical Modelling Using Empirical Validation. *Nanomaterials* **2021**, *11*, 3284. [CrossRef]
69. Lutter-Günther, M.; Bröker, M.; Mayer, T.; Lizak, S.; Seidel, C.; Reinhart, G. Spatter formation during laser beam melting of AlSi10Mg and effects on powder quality. *Procedia CIRP* **2018**, *74*, 33–38. [CrossRef]
70. Andani, M.T.; Dehghani, R.; Karamooz-Ravari, M.R.; Mirzaeifar, R.; Ni, J. A study on the effect of energy input on spatter particles creationduring selective laser melting process. *Addit. Manuf.* **2018**, *20*, 33–43. [CrossRef]
71. Panwisawas, C.; Sovani, Y.; Turner, R.P.; Brooks, J.W.; Basoalto, H.C.; Choquet, I. Modelling of thermal fluid dynamics for fusion welding. *J. Mater. Process. Technol.* **2018**, *252*, 176–182. [CrossRef]
72. Volpp, J. Spattering effects during selective laser melting. *J. Laser Appl.* **2020**, *32*, 022023. [CrossRef]
73. Louvis, E.; Fox, P.; Sutcliffe, C.J. Selective laser melting of aluminium components. *J. Mater. Process. Technol.* **2011**, *211*, 275–284. [CrossRef]
74. Yu, H.; Hayashi, S.; Kakehi, K.; Kuo, Y.L. Study of Formed Oxides in IN718 Alloy during the Fabrication by Selective Laser Melting and Electron Beam Melting. *Metals* **2019**, *9*, 19. [CrossRef]
75. Salehi, M.; Maleksaeedi, S.; Farnoush, H.; Sharon, N.M.L.; Meenashisundaram, G.K.; Gupta, M. An investigation into interaction between magnesium powder and Ar gas: Implications for selective laser melting of magnesium. *Powder Technol.* **2018**, *333*, 252–261. [CrossRef]
76. Ghasemi, A.; Fereiduni, E.; Balbaa, M.; Jadhav, S.D.; Elbestawi, M.; Habibi, S. Influence of alloying elements on laser powder bed fusion processability of aluminum: A new insight into the oxidation tendency. *Addit. Manuf.* **2021**, *46*, 102145. [CrossRef]
77. Zhang, W.; Wang, L.; Feng, Z.; Chen, Y. Research progress on selective laser melting (SLM) of magnesium alloys: A review. *Optik* **2020**, *207*, 163842. [CrossRef]
78. Cao, L.; Chen, S.; Wei, M.; Guo, Q.; Liang, J.; Liu, C.; Wang, M. Effect of laser energy density on defects behavior of direct laser depositing 24CrNiMo alloy steel. *Opt. Laser Technol.* **2019**, *111*, 541–553. [CrossRef]
79. Shi, X.; Yan, C.; Feng, W.; Zhang, Y.; Leng, Z. Effect of high layer thickness on surface quality and defect behavior of Ti-6Al-4V fabricated by selective laser melting. *Opt. Laser Technol.* **2020**, *132*, 106471. [CrossRef]
80. King, W.E.; Barth, H.D.; Castillo, V.M.; Gallegos, G.F.; Gibbs, J.W.; Hahn, D.E.; Kamath, C.; Rubenchik, A.M. Observation of keyhole-mode laser melting in laser powder-bedfusion additive manufacturing. *J. Mater. Process. Technol.* **2014**, *214*, 2915–2925. [CrossRef]
81. Patel, S.; Vlasea, M. Melting modes in laser powder bed fusion. *Materialia* **2020**, *9*, 100591. [CrossRef]
82. Chen, H.; Yan, W. Spattering and denudation in laser powder bed fusion process: Multiphase flow modelling. *Acta Mater.* **2020**, *196*, 154–167. [CrossRef]
83. Carluccio, D.; Demir, A.G.; Caprio, L.; Previtali, B.; Bermingham, M.J.; Dargusch, M.S. The influence of laser processing parameters on the densification and surface morphology of pure Fe and Fe-35Mn scaffolds produced by selective laser melting. *J. Manuf. Process.* **2019**, *40*, 113–121. [CrossRef]

84. Nadammal, N.; Cabeza, S.; Mishurova, T.; Thiede, T.; Kromma, A.; Seyfert, C.; Farahbod, L.; Haberland, C.; Schneider, J.A.; Portella, P.D. Giovanni Bruno, Effect of hatch length on the development of microstructure, texture and residual stresses in selective laser melted superalloy Inconel 718. *Mater. Des.* **2017**, *134*, 139–150. [CrossRef]

85. Zhang, X.; Liu, B.; Zhou, X.; Wang, J.; Hashimoto, T.; Luo, C.; Sun, Z.; Tang, Z.; Lu, F. Laser welding introduced segregation and its influence on the corrosion behaviour of Al-Cu-Li alloy. *Corros. Sci.* **2018**, *135*, 177–191. [CrossRef]

86. Farhad, A.; Mohammadi, Z. Calcium hydroxide: A review. *Int. Dent. J.* **2005**, *55*, 293–301. [CrossRef] [PubMed]

87. Laursen, C.M.; DeJong, S.A.; Dickens, S.M.; Exil, A.N.; Susan, D.F.; Carroll, J.D. Relationship between ductility and the porosity of additively manufactured AlSi10Mg. *Mater. Sci. Eng. A* **2020**, *795*, 139922. [CrossRef]

88. Trapp, J.; Rubenchik, A.M.; Guss, G.; Matthews, M.J. In situ absorptivity measurements of metallic powders during laser powder-bedfusion additive manufacturing. *Appl. Mater. Today* **2017**, *9*, 341–349. [CrossRef]

89. Barin, I.; Knacke, O. *Thermochemical Properties of Inorganic Substances*; Springer: Berlin/Heidelberg, Germany, 1973.

90. Liu, T.; Yang, L.J.; Wei, H.L.; Qiu, W.C.; Debroy, T. Composition Change of Stainless Steels during Keyhole Mode Laser Welding. *Weld. J.* **2017**, *96*, 258–270.

91. Zhao, Y.; Aoyagi, K.; Yamanaka, K.; Chiba, A. Role of operating and environmental conditions in determining molten pool dynamics during electron beam melting and selective laser melting. *Addit. Manuf.* **2020**, *36*, 101559. [CrossRef]

92. Qiu, C.; Panwisawas, C.; Ward, M.; Basoalto, H.C.; Brooks, J.W.; Attallah, M.M. On the role of melt flow into the surface structure and porosity development during selective laser melting. *Acta Mater.* **2015**, *96*, 72–79. [CrossRef]

93. Le, K.Q.; Tang, C.; Wong, C.H. On the study of keyhole-mode melting in selective laser melting process. *Int. J. Therm. Sci.* **2019**, *145*, 105992. [CrossRef]

94. Wu, Y.C.; San, C.H.; Chang, C.H.; Lin, H.J.; Marwan, R.; Baba, S.; Hwang, W.S. Numerical modeling of melt-pool behavior in selective laser melting with random powder distribution and experimental validation. *J. Mater. Process. Technol.* **2018**, *254*, 72–78. [CrossRef]

95. Ozawa, S.; Morohoshi, K.; Hibiya, T.; Fukuyama, H. Influence of oxygen partial pressure on surface tension of molten silver. *J. Appl. Phys.* **2010**, *107*, 014910. [CrossRef]

96. Ozawa, S.; Morohoshi, K.; Hibiya, T.; Fukuyama, H. Influence of oxygen partial pressure on surface tension and its temperature coefficient of molten iron. *J. Appl. Phys.* **2011**, *109*, 014902. [CrossRef]

97. Egry, I.; Ricci, E.; Novakovic, R.; Ozawa, S. Surface tension of liquid metals and alloys—Recent developments. *Adv. Colloid Interface Sci.* **2010**, *159*, 198–212. [CrossRef]

98. Tanaka, T.; Hack, K.; Hara, S. Use of Thermodynamic Data to Determine Surface Tension and Viscosity of Metallic Alloys. *MRS Bull.* **1999**, *24*, 45–51. [CrossRef]

99. Lu, H.M.; Jiang, Q. Surface Tension and Its Temperature Coefficient for Liquid Metals. *J. Phys. Chem. B* **2005**, *109*, 15463–15468. [CrossRef]

100. Shannon, G.; White, L.; Sridhar, S. Modeling inclusion approach to the steel/slag interface. *Mater. Sci. Eng. A* **2008**, *495*, 310–315. [CrossRef]

101. Thelwlis, C.; Milner, D.R. Inclusion Formation in Arc Welding. *Weld. Res. Suppl.* **1977**, *56*, S281–S288.

102. Baroud, C.N. Thermocapillarity. In *Encyclopedia of Microfluidics and Nanofluidics*; Li, D., Ed.; Springer: Boston, MA, USA, 2013.

103. Zhang, C.; Gao, M.; Wang, D.; Yin, J.; Zeng, X. Relationship between pool characteristic and weld porosity in laser arc hybrid welding of AA6082 aluminum alloy. *J. Mater. Process. Technol.* **2017**, *240*, 217–222. [CrossRef]

104. Hibiya, T.; Ozawa, S. Effect of oxygen partial pressure on the marangoni flow of molten metals. *Cryst. Res. Technol.* **2013**, *48*, 208–213. [CrossRef]

105. Cho, D.W.; Park, Y.D.; Cheepu, M. Numerical simulation of slag movement from Marangoni flow for GMAW with computational fluid dynamics. *Int. Commun. Heat Mass Transf.* **2021**, *125*, 105243. [CrossRef]

106. Le, T.-N.; Lo, Y.-L. Effects of sulfur concentration and Marangoni convection on melt-pool formation in transition mode of selective laser melting process. *Mater. Des.* **2019**, *179*, 107866. [CrossRef]

107. Fabbro, R. Melt pool and keyhole behaviour analysis for deep penetration laser welding. *J. Phys. D Appl. Phys.* **2010**, *43*, 445501. [CrossRef]

108. Chiang, W.C.; Tseng, I.-S.; Møller, P.; Hilbert, L.R.; Nielsen, T.T.; Wu, J.K. Influence of silver additions to type 316 stainless steels on bacterial inhibition, mechanical properties, and corrosion resistance. *Mater. Chem. Phys.* **2010**, *119*, 123–130. [CrossRef]

109. Gosheger, G.; Hardes, J.; Ahrens, H.; Streitburger, A.; Buerger, H.; Erren, M.; Gunsel, A.; Kemper, F.H.; Winkelmann, W.; Eiff, C. Silver-coated megaendoprostheses in a rabbit model—An analysis of the infection rate and toxicological side effects. *Biomaterials* **2004**, *25*, 5547–5556. [CrossRef]

110. Fikri, Ş. Practical reduction of manganese oxide. *J. Chem. Technol. Appl.* **2017**, *1*, 1–2. [CrossRef]

111. Riboud, P.V.; Muan, A. Melting Relations of CaO-Manganese Oxide and MgO-Manganese Oxide Mixtures in Air. *J. Am. Ceram. Soc.-Riboud Muan* **2006**, *46*, 33–36. [CrossRef]

112. Ahsan, R.U.; Cheepu, M.; Ashiri, R.; Kim, T.H.; Jeong, C.; Park, Y.D. Mechanisms of weld pool flow and slag formation location in cold metal transfer (CMT) gas metal arc welding (GMAW). *Weld. World* **2017**, *61*, 1275–1285. [CrossRef]

113. Yusop, A.H.; Bakir, A.A.; Shaharom, N.A.; Kadir, M.R.A.; Hermawan, H. Porous BiodegradableMetals for Hard Tissue Scaffolds: A Review. *Int. J. Biomater.* **2012**, *2012*, 641430. [CrossRef]

Modification of Iron with Degradable Silver Phases Processed via Laser Beam Melting for Implants with Adapted Degradation Rate

Jan Tobias Krüger,* Kay-Peter Hoyer, Anatolii Andreiev, Mirko Schaper, and Carolin Zinn

In medical technology, implants are used to improve the quality of patients' lives. The development of materials with adapted properties can further increase the benefit of implants. If implants are only needed temporarily, biodegradable materials are beneficial. In this context, iron-based materials are promising due to their biocompatibility and mechanical properties, but the degradation rate needs to be accelerated. Apart from alloying, the creation of noble phases to cause anodic dissolution of the iron-based matrix is promising. Due to its high electrochemical potential, immiscibility with iron, biocompatibility, and antibacterial properties, silver is suited for the creation of such phases. A suitable technology for processing immiscible material combinations is powder-bed-based procedure like laser beam melting. This procedure offers short exposure times to high temperatures and therefore a limited time for diffusion of alloying elements. As the silver phases remain after the dissolution of the iron matrix, a modification is needed to ensure their degradability. Following this strategy, pure iron with 5 wt% of a degradable silver–calcium–lanthanum alloy is processed via laser beam melting. Investigation of the microstructure yields achievement of the intended microstructure and long-term degradation tests indicates an impact on the degradation, but no increased degradation rate.

1. Introduction

An increasing number of diseases and injuries are treated with implants and improve the quality of the patient's life.[1–3] Therefore, implants and materials with adjusted properties for each application are required.[2–5] Materials fulfill the central requirement of biocompatibility if no adverse consequences like cancer, toxicity, allergic reaction, inflammatory, etc. occur. The most used implant materials are biocompatible due to inert behavior. But the acceptance of noninert properties and targeted degradation for self-dissolving implants and the release of substances with, e.g., an antibacterial effect from the implant enable the creation of implants with beneficial features.[2–4,6] Accordingly, the current research aims to exploit the potential of materials, e.g., to create antibacterial properties preventing implant-related infections, reduce stress-shielding by adaption of stiffness or bioresorbable implants. For many applications, the supportive function of implants is only required for a certain time, e.g., the fixture of bone fracture or treatment of arterial sclerosis with a cardiovascular stent. These implants can cause adverse consequences if they remain in the body and even the removal is connected to risks and patient burden. Thus, bioresorbable implants are highly promising to overcome these limitations.[1,5,7–11] A few degradable materials based on, e.g., polymers and magnesium (Mg) are already successfully used in clinical applications. But these materials do not fulfill the requirements of all possible applications, e.g., due to high degradation rates and low mechanical strengths. Accordingly, further materials based on, e.g., iron (Fe), zinc (Zn), and molybdenum (Mo) are discussed as bioresorbable materials.[2,4,7,8,12,13]

Following the potential of the application of Fe-based degradable materials and challenges addressed in this study are presented. The aspects of biocompatibility, degradation behavior, and the strategy applied in this study to adapt the degradation behavior are addressed. Iron is highly interesting for the development of degradable materials as it is characterized by suited mechanical properties, reduced degradation rate compared to Mg-based materials, and good biocompatibility.[7,12,14,15–17]

J. T. Krüger, K.-P. Hoyer, A. Andreiev, M. Schaper
Chair of Materials Science
Paderborn University
Mersinweg 7, 33100 Paderborn, Germany
E-mail: krueger@lwk.upb.de

J. T. Krüger, K.-P. Hoyer, A. Andreiev, M. Schaper
DMRC - Direct Manufacturing Research Center
Paderborn University
Mersinweg 3, 33100 Paderborn, Germany

C. Zinn
Chair of Materials Science and Testing
University of Siegen
Paul-Bonatz-Straße 9-11, 57076 Siegen, Germany

The ORCID identification number(s) for the author(s) of this article can be found under https://doi.org/10.1002/adem.202201008.

DOI: 10.1002/adem.202201008

As an uptake of up to 9.8 g per year of Fe is necessary to ensure a sufficient supply of the body with trace elements, the release of, e.g., 2 g Fe from an osteosynthesis plate over a few months is uncritical.[18,19] However, pure Fe is not suited and has to be modified to obtain mechanical properties comparable to conventional implants and increase the insufficient degradation rate.[5,7,20,21] Thus, alloying is necessary. Manganese (Mn) is promising as it increases the degradation due to the reduction of the electrochemical potential and improves the mechanical properties, e.g., ductility and strength. Thereby, relatively large amounts of up to 35 wt% Mn are necessary to reach the desired properties.[14–17,21–23] Although Mn is a necessary trace element, it is critically discussed regarding biocompatibility because Mn causes toxicity against the nervous system, lung, cardiovascular system, or liver. Thus, the use of pure Fe instead of Fe–Mn alloy contributes to improved biocompatibility and pure Fe is selected to be investigated.[19,24–26] The strategy presented in this study might enable the waiver of Mn; meanwhile, a satisfactory degradation rate is achieved. However, an adaption of the mechanical properties by alloying with low amounts of, e.g., carbon (C) should be possible without expectable significant influence regarding biocompatibility and degradability.

Nevertheless, the concern of an insufficient degradation rate still has to be addressed. Therefore, the creation of phases with different electrochemical potentials is promising as areas with lower potential are increasingly attacked due to anodic dissolution. As pure Fe or low alloyed Fe is promising regarding biocompatibility and mechanical properties, the creation of phases with high electrochemical potential within a Fe matrix is aimed.[27–31] As silver (Ag) is insoluble with Fe in the solid as well as in the liquid state, it enables the existence of Ag phases with high electrochemical potential in an unaffected Fe matrix. Enhanced degradation due to the addition of Ag is confirmed by various studies.[14,27,28,31,32] As a good compromise between biocompatibility, mechanical properties, and increased degradation, an amount of 5 wt% Ag is selected for the present study.[27,28,33] Biocompatibility of such small amounts of Ag can be assumed as it is utilized in medicine for a long time, e.g., for the treatment of burns. Ag acts antibacterial and a suited release of Ag ions can prevent implant-related infections.[34–37]

However, pure Ag would remain after the degradation of the matrix material (Fe). These remaining Ag particles can cause adverse consequences, e.g., due to the distribution in the human organism and the consequent release of Ag in critical areas like the brain.[38–41] Thus, Ag has to be modified by alloying to create a degradable Ag alloy to fulfill the aim of complete bioresorbability. The electrochemical potential of such alloy is still high enough to enhance the anodic dissolution of the surrounding matrix material. In contrast to pure Ag, the remaining particles of the degradable Ag alloy dissolve completely after the dissolution of the Fe-based matrix and thus, the implants dissolve residue-free.[42–45] Biocompatibility of such alloys can be assumed due to the descriptions and investigations undertaken in previous studies.[42,44] For the present study, an Ag alloy with 12 wt% Ca and 5 wt% La (AgCaLa) is selected.[42] Pure Fe is modified with 5 wt% of this alloy.

For designing such material, the biocompatibility and degradability of corrosion products that are formed in the complex environment of the living organism have to be considered.

Here, investigations of in vivo degradation of Fe reveal good biocompatibility.[17,20,24,46–51] However, the complete removal and excretion of corrosion products are arguable as hardly dissolvable degradation products are developed on the implantation site causing discoloration of surrounding tissue.[47,48,50,51] Nevertheless, complete removal of residues by macrophages and the lymphatic system can be assumed, as degradation products transported to the lymph nodes indicating the excretion are observed 52 months after implantation of Fe stents.[14,17,48]

However, the degradation of Fe-based alloys is complex and strongly dependent on the composition of the electrolyte, the presence of, e.g., proteins, the amount of dissolved O which is necessary for the cathodic reaction as well as the pH value which is decisive for the solubility of degradation products.[14,22,30,50,52,53] In the human body, many different elements and chemical compounds prevail causing the formation of various corrosion products, consequently resulting in a multilayer structure that further influences the degradation.[52,54,55] For example, compounds of Ca with phosphor (P) on the top of the corrosion products can create an inhibiting layer as they suppress the contact with the electrolyte and the diffusion of O.[15,49,54,56–58] Thus, the formation of such inhibiting layers must be prevented. The addition of noble phases can contribute to the suppression of the blocking effect of such layers. Wang et al.[52] suggest a change in the structure or detachment of such inhibiting layers due to the addition of Ag. Accordingly, for in vivo investigations, an increased degradation is reported for Fe-based materials modified with Ag.[14,59] It has to be mentioned that even the opposite effect is reported.[47] Nevertheless, an increased degradation due to the addition of Ag is expectable; nonetheless, the material has to be modified to obtain a reliable and suited degradation of the material.

From the foregoing arises the need for a Fe-based matrix containing dispersed phases of a degradable Ag alloy because Fe modified with degradable Ag phases is promising to create a bioresorbable material. But the processing of such immiscible material combinations is challenging. Powder-bed-based processes are suitable for the fabrication of such materials as they enable the mechanical mixing of the components.[60–64] Laser beam melting (LBM) is characterized by small melt pools with strong melt flow as well as rapid melting and solidification which enables the processing of such materials.[27,65–68] Following crucial aspects of the LBM process, on which the conducted research is based, are presented. The short exposure time to high temperatures during LBM is beneficial for the processing of degradable Ag alloys as the time for diffusion of alloying elements between both components is limited. The diffusion of alloying elements between immiscible components can be distinctive. Even for processing of pure Ag with FeMn via LBM, an uptake of Mn by the Ag can be observed.[68–70] But the uptake of Ca and La from the AgCaLa alloy by the FeMn matrix is uncritical as both elements can be used as deoxidizers.[71,72] Thus, diffusion of Ca and La is acceptable as long as enough Ca and La remain in the Ag phases to still obtain degradable Ag phases. However, the process and material must be adapted to achieve a suited chemical composition of the Ag phases as well as of the matrix material. Thus, the use of pure Fe instead of FeMn might be beneficial as Ag cannot take up any Mn from the matrix material.

Apart from the chemical composition, the morphology of the Ag phases is decisive as, e.g., nanoparticular Ag is critically discussed and should be avoided.[67,73] Otherwise, an even distribution of small Ag phases is necessary to obtain homogenous degradation and reduce impairments of mechanical properties. The Ag phases do not contribute to load capacity as no load-carrying connection between the Ag and the matrix exists.[68] Thus, the possibility for targeted adaption of the morphology of Ag phases is crucial for the successful application of LBM to obtain a suited morphology. A previous study by Krüger et al.[68] proves the possibility to influence the morphology by the adaption of the LBM process parameters. For the targeted adaption, a thorough understanding of the physical effects and mechanisms of the formation of the Ag phases is essential. Thus, the present work aims to provide a fundamental understanding of the formation of Ag phases by the reflection of the resulting microstructure in conjunction with information from the literature. Simultaneously, the study aims to evaluate the suitability of LBM to process Fe with a degradable Ag alloy and, in general, the suitability of the processed material as a degradable alloy.

Gaining knowledge of the LBM process and the influencing physical effects is a subject of the present research. Models are developed based on knowledge of physical effects combined with results from the monitoring of the melt pool and resulting microstructure.[74–78] These results can be used to understand the interaction of two immiscible materials during simultaneous processing via LBM.

During the sequential proceeding of LBM, physical effects caused by laser radiation result in the successive formation of bulk material while the laser moves along the melt track. The decisive effects are the absorption of laser radiation, melting, consolidation of the melt, melt flow, heat dissipation, and solidification.[74–76,78] These effects are related to each other and cause further effects like spattering.[74,78–80] For instance, the amount of absorbed laser energy during the LBM process depends not only on the absorptivity of the materials as a reflection within the powder bed or an eventually formed keyhole, and the deflection by the vapor plume, etc. are also decisive.[81–83] A keyhole develops if a threshold value of energy introduction is exceeded, and the recoil pressure of vaporized material is high enough to create a deep depression.[84–86] As the laser is reflected multiple times within the keyhole, the energy absorption increases due to the formation of a keyhole.[84,87,88] The threshold value of the necessary energy input depends on the chemical composition of the material as alloying elements with high vapor pressure increase the recoil pressure and ultimately reduce the necessary energy input.[85–88] The presence of a keyhole defines the keyhole welding mode. If no keyhole occurs, the energy is mainly transferred by conduction to deeper parts of the material and the so-called conduction mode is present. Between both modes, a transition mode with partitions of both is defined.[75,82,89,90]

The present welding mode influences the shape of the melt pool as well as the melt flow and thus, the formation of the Ag phases. As the first part of the melt pool in front of and in the interaction zone with the laser is mainly determined by the vapor pressure which acts equally to both components, an emerging of Ag occurs.[68,75,76] Apart from the vapor pressure, the surface tension is the decisive driving force for melt flow and determines the shape of the melt pool. Especially in the part of the melt pool behind the interaction zone with the laser, surface tension and resulting effects like the Marangoni flow are the decisive factors.[75–77,81,82,91] As the shape of Ag phases is fixed in the moment of solidification and the shape in the moment of solidification results from previous melt flow, surface tension is even crucial for the morphology of the Ag phases.[68] Thus, surface tension and surface tension-driven effects have to be considered to develop a model of the formation of the Ag phases. Krüger et al.[68] observed intensive agglomeration of Ag on the top of an as-built top surface with increased thickness of the Ag layer between adjacent melt pools explainable by surface tension-driven effects: The agglomeration of Ag reduces the surface between FeMn and Ag and thus, the required surface energy. The position of Ag on the top of FeMn also reduces the overall needed surface energy, as the surface tension of the Ag phase is lower than that of the FeMn phase. Thus, a top surface of Ag requires less energy than a surface of FeMn. Accordingly, additional surface energy would be required to remove Ag from the top surface as an additional surface between Ag and FeMn would be created and additional surface energy to create a top surface of FeMn would be necessary. At last, a surface tension-driven outward-directed vortex of the FeMn under the Ag explains the entrainment of Ag to the sides of the melt pool. Thus, Ag phases in the bulk material originate from the top layer. Parts of the Ag layer between adjacent melt tracks remain if they are deep enough to be not remelted during the generation of the next layer.

However, further effects like spattering influence the composition and structure of the mixed material as spattering depends on material properties as well as on the process conditions.[68,74,79,92,93] As ejections are caused mainly by recoil pressure and Marangoni forces resulting from surface tension, the influence of material properties on the spattering is conclusive.[74,79,83,94] Thus, one of the immiscible components might be more likely to be ejected by spattering and the proportion of both materials might be influenced.

Beyond the physical properties, the chemical properties can influence the LBM process. Typically, small amounts of O remain in the build chamber; therefore, an interaction between the alloying elements and O can occur.[95–97] As the high temperature during LBM increases the reactivity, chemical compounds may be formed that have a detrimental effect on the material properties.[95,96,98,99] Slag from the top can be incorporated into the bulk material and can negatively affect the mechanical properties.[96,100,101]

Thus, a lot of influencing effects have to be considered to explain the morphology and chemical composition of the Ag phases. The results gained within this study regarding the interaction of immiscible phases during LBM and the suitability of the investigated material as bioresorbable material enable further targeted adaption.

Summarizing, Fe-based materials are promising as bioresorbable materials, but the degradation rate must be increased. The applied strategy is based on the acceleration of dissolution by insertion of degradable Ag phases causing anodic dissolution of Fe. The use of a degradable AgCaLa alloy should enable the dissolution of AgCaLa phases after the dissolution of the matrix. Biocompatibility of Fe modified with AgCaLa can be assumed.

The processing of such immiscible material combinations is possible via LBM and the study reports first about the processing and microstructure of such material. These investigations are necessary to understand and adapt the LBM process influenced by effects such as the diffusion of alloying elements, melt flow, spattering, etc. In the second part of the present study, the impact of the AgCaLa phases on the degradation behavior is extensively investigated to evaluate the suitability of the newly developed material for the intended application.

2. Experimental Section

To achieve good comparability of the results, the experimental procedure is almost identical to the one performed by Krüger et al.[68] Even the parameters applied for LBM are identical. As stated in the introduction, the modification of Fe with 5 wt% of a degradable Ag alloy is addressed. Therefore, Fe is processed with 5 wt% of an AgCaLa alloy as well as with 5 wt% Ag and pure Fe without any modification as a reference. Following FeAgCaLa describes the LBM-processed mixture of Fe with AgCaLa whereas FeAg stands for the LBM-processed mixture of Fe with pure Ag. AgCaLa and Ag describe the added degradable Ag alloy and pure Ag and the phases of those materials in the Fe matrix.

Fe and Ag powders (TLS Technik GmbH & Co. Spezialpulver KG, Bitterfeld-Wolfen, Germany) are produced via argon-gas-atomization with a nominal particle size of 15–53 μm and a nominal Fe content of 99.8% and a particle size of 20–63 μm and an Ag content of 99.5%.

To produce the AgCaLa powder (Nanoval GmbH & Co. KG, Berlin, Germany) via argon-gas-atomization, an AgCaLa alloy containing 12.9 wt% Ca and 5.7 wt% La (measured via inductively coupled optical emission spectroscopy, ICP-OES) is used.

The particle size distribution (PSD) is measured using a Mastersizer 2000 (Malvern Panalytical GmbH, Kassel, Germany). The absorption of the powder bed is determined by utilizing a HL-2000-HP-FHSA (Ocean Optics, Inc., Orlando, Florida, USA).

The powders are mixed for 30 min at 60 rpm with a drum hoop mixer with half-filled powder containers. Due to vacuum drying, the humidity of the powder is less than 5%. The powders are processed with an SLM 280^{HL} (SLM Solutions Group AG, Lübeck, Germany) in an argon atmosphere (4.6) with an oxygen level of less than 0.03%. For processing a layer thickness of 50 μm in combination with a scanning velocity of 750 mm s^{-1}, a laser power of 280 W, a hatching distance of 110 μm, and a rotation of 67° between the scanning direction of the following layers are applied. All samples (cubes: $11 \times 13 \times 8$ mm^3) of each material are produced within one build job.

The preparation of the samples for microstructural investigations starts with a stepwise grinding process up to a final step with P4000. At least 1 mm is removed to prevent edge effects. The subsequent vibration polishing for 12 h is performed with a mixture of 50% corrosion inhibiting lubricant (CUTLUB (Diluted, 3%), Cloeren Technology GmbH, Wegberg, Germany) and silica polishing suspension (grit size: 50 nm, Cloeren Technology GmbH, Wegberg, Germany). For etching, the samples are exposed to 0.5% HNO$_3$ acid (nitric acid).

Light microscopy (LM) investigations are undertaken with an Axiophot (Carl Zeiss AG, Oberkochen, Germany).

For investigations with a field emission scanning electron microscopy (FE-SEM), a Zeiss Ultra Plus (Carl Zeiss AG, Oberkochen, Germany) with a secondary electron (SE) detector and a backscatter electron (BSE) detector is used. The FE-SEM is equipped with an electron backscatter diffraction (EBSD) detector DIGIVIEW 5 (AMETEK, Berwyn, PN, USA) for the investigation of grain structure and an energy-dispersive X-ray spectroscopy (EDS) detector Octane Pro (AMETEK, Berwyn, PN, USA) for the determination of the chemical composition. Visualized EBSD data (color-coded orientation plot superimposed with image quality) are presented without any cleaning. To gain an impression of the 3D distribution of the Ag phases, cuboids with a length of 4 mm are cut from the edge of the initial cuboids and investigated with a micro-CT SkyScan 1275 (Bruker Corporation, Billerica, Massachusetts, USA). The chemical composition of the AgCaLa phases within the Fe matrix is determined via wavelength dispersive X-ray (WDX) analysis by using a Jeol JXA-8100 (Jeol Denshi K.K., Akishima, Tokyo, Japan).

For the 56 days lasting static immersion test samples with a dimension of $10 \times 13 \times 2$ mm^3 are cut from the initial cuboids and ground up to a grid size of P4000. Holes with a diameter of 1 mm are drilled in the samples to hang up the samples with a polypropylene filament. Three identical samples are placed free hanging in a glass bottle with a plastic screw cap. In the caps, three holes with a diameter of 3 mm are drilled to fix the samples. Thus, the immersion solution can interact with the surrounding atmosphere influencing, e.g., the amount of dissolved O. As 100 mL of electrolyte is utilized, a ratio of 10.9 mL cm^{-2} is reached. For the immersion test, three different electrolytes (**Table 1**) to identify the influence of components, e.g. P, on the degradation behavior are selected. During the immersion test, the samples are inspected after 1, 3, 6, 10, and subsequently every 7 days. The Hanks' solution (HS) with Ca^{++} and Mg^{++} (Lonza Group AG, Basel, Swiss), Ringer's lactate solution (RL) (B. Braun SE, Melsungen, Germany), and the NaCl solution (NC) are changed at every inspection. For inspection, the samples are removed from the bottles and rinsed (slow movement in a bath) iteratively in Ringer's rinsing solution cleaner

Table 1. Composition of solutions utilized for static immersion test.

Components		Concentration [g L^{-1}]		
		HS	RL	NC
Calcium chloride dihydrate	CaCl$_2$ · 2H$_2$O	0.186	0.270	
Dextrose	C$_6$H$_{12}$O$_6$	1.000		
Magnesium sulfate heptahydrate	MgSO$_4$ · 7H$_2$O	0.200		
Potassium chloride	KCl	0.400	0.400	
Potassium phosphate monobasic anhydrous	KH$_2$PO$_4$	6.000E–02		
Sodium bicarbonate	NaHCO$_3$	0.350		
Sodium chloride	NaCl	8.000	6.000	9.000
Sodium phosphate dibasic-7-hydrate	Na$_2$HPO$_4$ · 7H$_2$O	9.000E–02		
Phenol red	C$_{19}$H$_{14}$O$_5$S	2.000E–02		
Sodium lactate	NaC$_3$H$_5$O$_3$		3.120	

(B. Braun SE, Melsungen, Germany) and acetone to absorb the water. Due to this treatment, only loose degradation products are removed. Complete removal of degradation products is not aimed, as this does not occur in the human body. Following, the samples are dried for 30 s with hot air and the weight is determined with a precision scale XP205 (Mettler-Toledo, LLC, Columbus, OH, USA) with an accuracy of 0.01 g. For investigations of the samples via SEM after the immersion test, the samples are additionally cleaned for 30 s in an acetone ultrasonic bath.

3. Results and Discussion

3.1. LBM Process and Resulting Microstructure

The raw material is characterized, as it influences the properties of parts produced via LBM. A good flowability can be assumed as all powders have a suited PSD and morphology as the particles are spherical and just a few satellites are observed (**Figure 1**a,c,d,e). Only the Ag powder shows a few more satellites. But as only 5 wt% of this powder is used in combination with Fe, a suitable flowability is postulated. Apart from the morphology, the chemical composition is decisive. As measured with ICP-OES, the AgCaLa powder contains 12.4 wt% Ca, 5 wt% La, 0.3 wt% O, and Ag (bal.). Thus, the chemical composition corresponds satisfactorily with the intended composition.

However, to obtain well-dispersed Ag phases in the final bulk material, a homogenous distribution of the particles in the powder bed is necessary. An even distribution of the different materials can be assumed, as the properties of the powders are comparable to the powders investigated by Krüger et al.[68] and they have proven an even distribution of Ag particles in the entire powder bed. The addition of Ag and AgCaLa slightly influences the absorption of laser radiation by the powder bed as the absorption of laser radiation is 61.5% for Fe, 59.5% for FeAg, and 61% for FeAgCaLa.

Welding spatters and particles with a size over 100 μm are sieved out after the LBM process and investigated as they provide information regarding the LBM process. Apart from the

enclosure of Fe particles by AgCaLa which is explainable due to the different melting points, irregularities of welding spatters resulting from the processing of Fe and FeAgCaLa are not observed. The welding spatters resulting from the processing of Fe with Ag contain significantly more than 5 wt% Ag (Figure 1b). Thus, a preferred ejection of Ag from the melt pool occurs.

The as-built top surface is characterized (**Figure 2**), as it corresponds to surfaces that are covered by fresh powder and remelted during the generation of each new layer. The surface of FeAg shows an increased amount of Ag compared to the added 5 wt% Ag, as significantly more than 5% of the top surface are covered by Ag (Figure 2a). The Ag phases are preferably aligned along the sides of single melt tracks, but even narrow alignments of Ag along the center of the melt track are observed (Figure 2a,c). As expected, Fe is detected aside from Ag and just small amounts of O can be determined.

Contrary, the surface of FeAgCaLa samples significantly differs (Figure 2b,d). The single melt tracks are not observable and larger irregularities like depressions or raises exist which indicate a more discontinuous LBM process. Furthermore, a significant enrichment of AgCaLa results in a nearly closed AgCaLa layer. Compared to the FeAg surface, the detection of less Fe underlines the distinct character of the AgCaLa layer.

However, aside from Ag, the alloying elements Ca and La are detected. In comparison to La which is mainly arranged in insular structures, Ca is, apart from some enrichments and depletions, homogeneously distributed. Moreover, a significant amount of inhomogeneously distributed O is detected. Thus, the formation of differently O-rich chemical compounds with the reactive alloying elements and a resulting layer of slag on the top are conclusive. As the SEM-EDS measurements reflect the surface near chemical composition, the homogenously distributed amounts of Ca together with the detected Ag in the same area might indicate that Ca remains as intended in the AgCaLa phases. Furthermore, the lamellar structures observed on the top of FeAgCaLa samples (Figure 2d) support the hypothesis that the AgCaLa alloy remains with properties comparable to

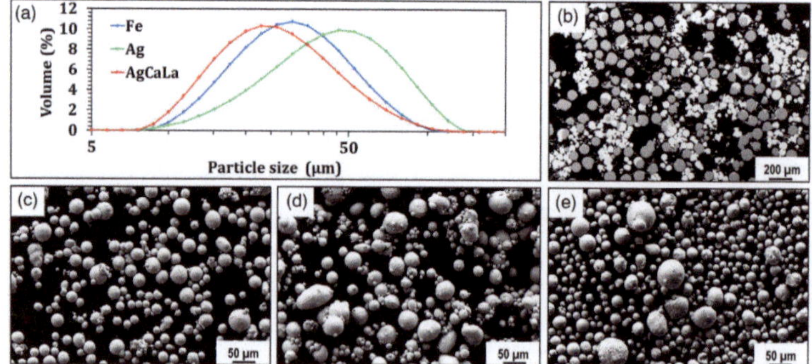

Figure 1. a) PSD of powder raw material; SEM–SE: c) Fe powder, d) Ag powder, e) AgCaLa powder; b) SEM–SE: cross section of oversize particles of FeAg sieved out after the LBM process, particles with a bright appearance: Ag.

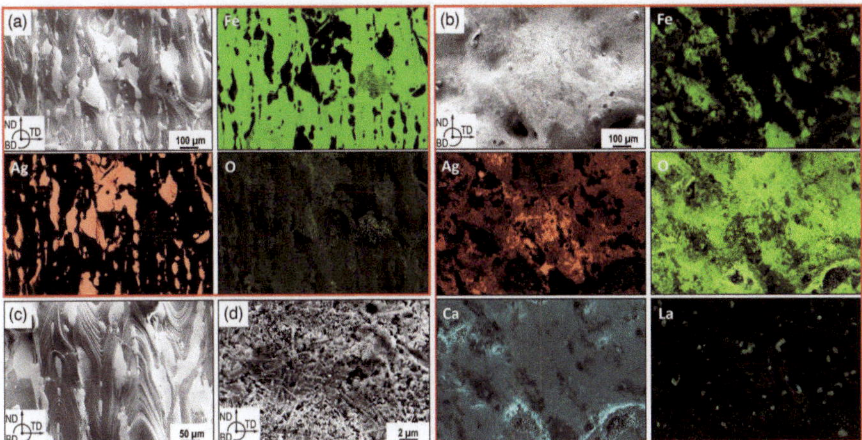

Figure 2. As-built top surfaces: a) SEM-SE with the corresponding distribution of elements measured via SEM–EDS: FeAg, Ag aligned along the melt tracks; b) SEM–SE with the corresponding SEM–EDS: FeAgCaLa, enrichment of Ag on the top and formation of O-rich compounds; c) SEM–SE: FeAg; d) SEM–SE: FeAgCaLa, partially lamellar structure.

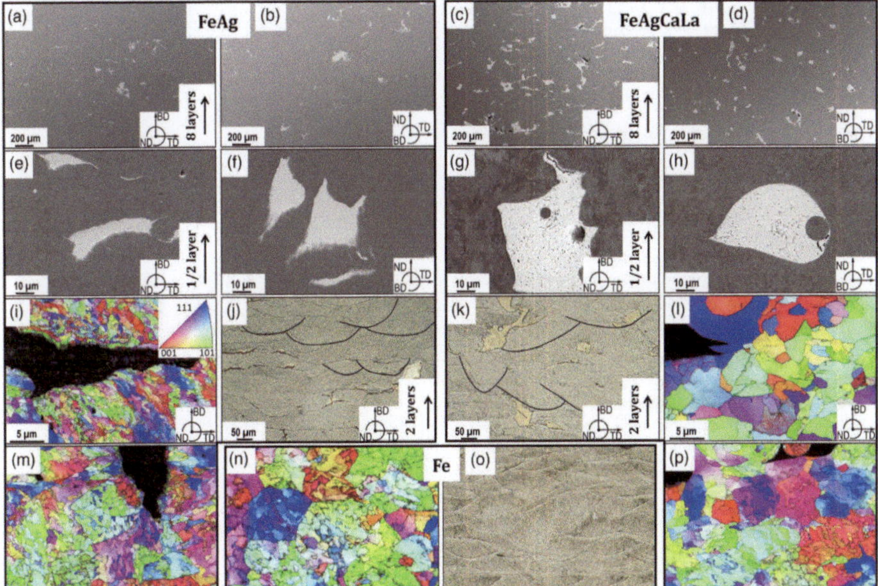

Figure 3. Microstructure and morphology of Ag phases: a,b,e,f) SEM-SE: FeAg, homogenous distributed Ag phases (bright areas) with narrow shape; c,d,g,h) SEM–SE: FeAgCaLa, homogenous distributed AgCaLa phases (bright areas), more AgCaLa than Ag in the bulk material; i,m) SEM-EBSD: FeAg (dark areas: Ag phases); j) LM: etched FeAg (bright areas: Ag phases); k) LM: etched FeAgCaLa (bright areas: AgCaLa phases); l,p) SEM-EBSD: FeAgCaLa (dark areas: AgCaLa phases); n) SEM-EBSD: Fe; o) LM: etched Fe.

conventional processed AgCaLa as such structures are also observed for a conventionally manufactured alloy.[42]

Differences between both material combinations are observed in the bulk material as well as on the top surface. Both materials contain relatively homogenously distributed Ag phases and AgCaLa phases in the bulk material (**Figure 3**a–h); the shape of the phases and the amount differ. FeAg (Figure 3a,b) contains narrow-shaped Ag phases which occupy a smaller part of the cross section compared to the AgCaLa phases in FeAgCaLa (Figure 3c,d). The shape of AgCaLa phases is more suitable for the intended application because the compact round-shaped AgCaLa phases affect the mechanical properties less due to the lower notch effect compared to the narrow shape of Ag phases with additionally increased impairment of the load-bearing cross section. However, this hypothesis must be confirmed by further investigations which would go beyond the scope of this study.

More AgCaLa is incorporated into the bulk material than Ag, although for both materials a proportion of 5 wt% is used. But it has to be considered that the density of AgCaLa is ≈42% lower than that of Ag due to the lower density of the alloying elements and thus, more AgCaLa is expectable. However, the differences are as distinctive as the influence of further effects is conclusive. Thus, a reduction of the Ag content for FeAg compared to the added 5 wt% has to be assumed. This is conclusive, as the welding spatters of FeAg contain an increased amount of Ag (Figure 1b) and the spatters of FeAgCaLa do not. Thus, some of the Ag is removed by welding spatters whereas most of the AgCaLa is incorporated into the bulk material. However, the quantity of Ag phases and AgCaLa phases seems not to differ significantly, but the AgCaLa phases have a less narrow, compact shape.

FeAg has a high density and no significant pores are visible (Figure 3a,b), whereas within FeAgCaLa some pores preferential in conjunction with the Ag phases exist (Figure 3c,d). Under consideration of the irregular structure of the top surface of FeAgCaLa (Figure 2b) this is conclusive. Furthermore, due to the slag observed on the top, the incorporation of oxides into the bulk material is expectable, but significant inclusions are not observed. Thus, the slag is removed by, e.g., welding spatters, transferred to the top of a newly generated layer, and less is incorporated.[92,102]

Accordingly, the addition of 12 wt% Ca and 5 wt% La to Ag, which corresponds to an amount of only 0.85 wt% referred to the Fe mixed material, influences the LBM process significantly. Slight differences are observed regarding the shape of the melt pools and the grain structure (Figure 3j,k,o). The melt pools of all three investigated materials are flat, indicating the conduction welding mode.[79,82,90,103] The melt pools of FeAgCaLa seem to be larger and deeper and some show the typical shape resulting from transition welding mode (Figure 3k). Such shapes are observed, but less distinctive for Fe (Figure 3o).[79,82,90,103]

The EBSD analysis confirms the expected ferritic structure of the Fe matrix. Fe and FeAg are characterized by small grains with a lot of small-angle grain boundaries (Figure 3i,m,n). Lejček et al.[104] suggest that this structure results from the repeated transition between α and γ and the decay of the γ grains into smaller α subgrains. Furthermore, the grain structure indicates differences in the solidification process. The grains of FeAgCaLa

are bigger and do not show this distinctive structure of subgrains (Figure 3l) in the section parallel to the building direction. These observations are confirmed by the grain structure in the SEM-SE images (Figure 3e,g) where the gray discolorations of the matrix material correspond to the grains. The scale of these discolorations is increased for FeAgCaLa compared to FeAg. However, structures of subgrains are observed for FeAgCaLa in a section perpendicular to the building direction (Figure 3p). Thus, the solidification conditions between the materials differ, but even within the FeAgCaLa a differing proceeding of solidification has to be assumed. This hypothesis of varying conditions for FeAgCaLa is supported by the top surface, indicating a less stable melt pool as well (Figure 2b).

The investigations of the 3D structure of Ag and AgCaLa inside the Fe matrix confirm the impression gained from the 2D cross sections. The FeAgCaLa contains more AgCaLa than the FeAg contains Ag (**Figure 4**b,c). For FeAg, a fraction of 4.5 vol% is detected via CT, whereas for FeAgCaLa a fraction of 10.1 vol% is found. However, these values are probably too high as more AgCaLa is detected than has been added to the material as 5 wt% corresponds to 6.4 vol% AgCaLa. The same applies to FeAg, as 5 wt% Ag corresponds to 3.8 vol% Ag. The reason for the higher values is probably the selection of the threshold value for the separation between Fe and Ag. However, as more AgCaLa (4.4 vol%) is determined than expected, and only 0.7 vol% Ag more than expected is detected, a value significantly less than 4.5 vol% Ag and the increased ejection of Ag (Figure 1b) are still conclusive.

The different sizes of the Ag and AgCaLa phases observed in cross sections are only partially confirmed by the CT investigations, as the size distribution of the Ag and AgCaLa phases is almost similar (Figure 3a). Only in the section with the largest

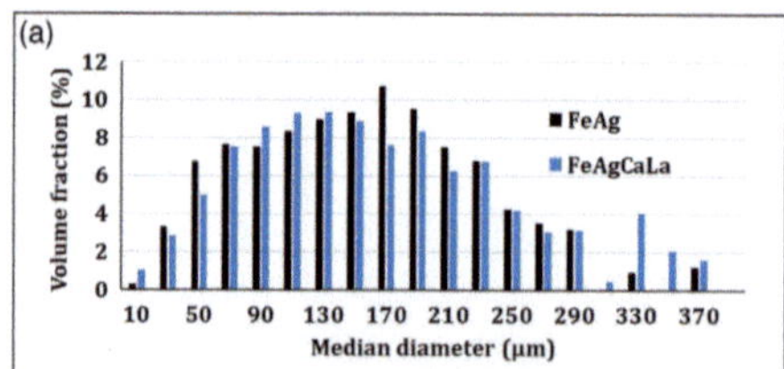

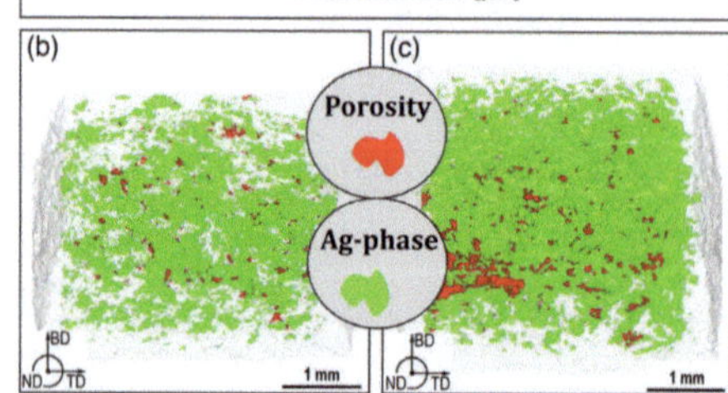

Figure 4. Distribution of Ag and AgCaLa phases three-dimensionally analyzed via CT: a) distribution of particle size; b) homogenous distribution of Ag in FeAg, a few pores; c) homogenous distribution of AgCaLa in FeAgCaLa, increased amount of AgCaLa and pores.

particle sizes of more than 290 µm, a slightly increased fraction of AgCaLa is detected. Thus, the AgCaLa phases are bigger, but the difference might be less distinctive than prevailed by the investigation of the cross sections.

3.2. Impact of Ag/AgCaLa on the LBM Process and Resulting Morphology

The addition of Ag and AgCaLa significantly influences the LBM process, and the morphology of the generated phases differs. As already described in the introduction, properties of the processed materials like vapor pressure and surface tension are crucial.[75,91] As the measured absorption coefficients slightly differ, the influence of the absorption can be neglected.

At first, the shape of the melt pools indicates the conduction mode[79,82,90,103] which is, in contrast to the transition mode, observed for the processing of FeMn with Ag as determined by Krüger et al.[68] As the LBM parameters are identical and the absorption of FeMn with Ag is only 5.5%[68] higher than for FeAgCaLa, another effect must be responsible for the changed welding mode. The vapor pressure of Fe is lower than the one of Mn and thus, the absence of depression and smaller melt pools for Fe compared to FeMn are conclusive.[85–87,105] As Ca has a significantly higher vapor pressure, even small amounts of Ca can lead to depression and increased absorption.[105,106] Thus, the influence of vapor pressure explains the increased and deeper melt pools for FeAgCaLa. Furthermore, the partial absence of subgrains for FeAgCaLa indicates changes in the solidification conditions which might be correlated to the changed shape of the melt pool.

The chemical reactions indicated by the slag contribute to the unstable melt pool during the processing of FeAgCaLa which is indicated by the top surface, pores, and differing grain structures. Chemical reactions provoke, e.g., release of additional energy or deflection by the vapor plume.[74,76,83,107] Moreover, the accumulation of Ag on the top contributes to an unstable melt pool. These varying influences are accumulated over a few layers and finally result in instability causing pores, etc.[75,68,91] Nevertheless, the matrix remains almost unchanged compared to the pure Fe as intended and overall suited morphology of AgCaLa is obtained. The distinctive Ag layer and compact-shaped Ag phases are comparable to the morphology observed for FeMn with Ag.[68] Thus, a similar formation of AgCaLa phases is conclusive and the possibility for the aimed adaption of the Ag phases compiled for FeMn with Ag can be adapted for FeAgCaLa.

In contrast, the narrow shape of Ag phases in FeAg differs from the shape of AgCaLa phases. To explain this shape, differences in surface tension have to be considered because the shape of Ag phases is defined in the backward part of the melt pool which is mainly determined by surface tension.[75–77,81,82,91] At first, a higher surface tension for Fe than for Ag is assumed.[54,68,108,109] As the surface tension of an alloy usually lies between the ones of the single elements and Ca and La have a lower surface tension than Ag, lower surface tension can be assumed for AgCaLa in comparison to Ag.[110–112] Further effects like O absorption could also have a decisive influence on the surface tension.[110,113] However, under the assumption of higher surface tension for Ag than for AgCaLa, a reduced difference

in surface tension between Fe and Ag, compared to Fe and AgCaLa, is conclusive. This reduces the necessary surface energy and the formation of narrow Ag phases with a large surface is simplified. In addition, further influences by viscosity and melting temperatures might be relevant. However, a small difference in surface tension reduces the driving force for the accumulation of Ag phases. Accordingly, the incorporation of Ag into the bulk material becomes easier as even the effects described in the introduction which keep Ag on the top are weaker. Thus, the shape of Ag phases indicates a small difference in surface tension between Ag and Fe and confirms the suggestions made before. The absence of a closed Ag layer on top of Fe supports the hypothesis of easier incorporation. Further effects like the changed welding mode, change of melt flow, and increased spattering of Ag contribute to the absence of an Ag layer. The material properties and resulting process characteristics are responsible for the increased spattering of Ag.[74,92,93]

As Ag on top is dragged along like slag with the underlying melt flow, the position of Ag enables conclusions regarding the underlying melt flow.[68,114,115] As Ag is increasingly observed at the sides of the melt pool, an outward-directed vortex, like for FeMn with Ag, is conclusive.[68] The narrow arrangements of Ag in the middle of the melt tracks indicate a more complex melt flow. As Ag remains at positions where Fe flows downward, a downward flow in the middle and on the sides of the melt pool is indicated. Thus, a complex melt flow with two vortexes on each half of the melt pool might exist.[116,117]

The melt pool of FeAg is more stable than the one of FeAgCaLa because the top surface is characterized by regular structures and fewer pores are observed. Due to the shape of Ag phases, the absence of the closed layer of Ag, and the suggestions concerning the melt flow, the formation of the Ag phases in Fe significantly differs from the formation of Ag phases in FeMn. But as pure Ag is not suited anyway due to its nondegradability, a deeper investigation of the formation of the Ag phases is not aimed. FeAg is the reference for the comparison with FeAgCaLa. Thus, a suitable morphology and distribution of the Ag and AgCaLa phases are observed for both materials; the different amounts and shapes of the Ag and AgCaLa phases have to be considered for the evaluation of the degradation behavior.

3.3. Microstructure and Suitability of AgCaLa Phase After Processing via LBM

The chemical composition and microstructure of the Ag and AgCaLa phases are decisive for their electrochemical potential and thus, the possibility to increase the degradation rate of the surrounding matrix. Furthermore, the degradation of the AgCaLa phases themselves after the dissolution of the matrix is influenced. Thus, the degradable AgCaLa alloy developed by Krüger et al.[42] should remain despite the processing via LBM with unchanged properties compared to the conventionally processed material.

However, the incorporation of slag has to be assumed, as detailed BSE investigations of the AgCaLa reveal the uptake of small particles of slag as several dark inclusions are observed (**Figure 5**a). The samples for the WDX measurement are ground

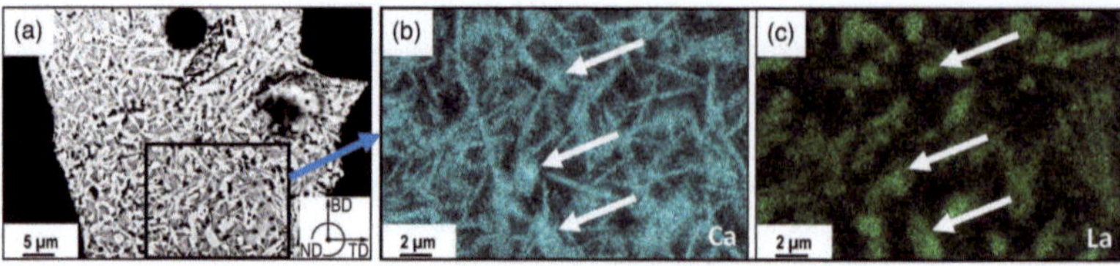

Figure 5. Microstructure of AgCaLa phases in Fe matrix: a) SEM-BSE with b,c) corresponding distribution of alloying elements detected via SEM-EDS, lamellar structure, and insular enrichment of La.

with ethanol up to a grid size of P2500 and subsequently polished utilizing a water-free suspension with a grid size of 3 μm as the water-based preparation might influence the chemical composition due to the reactive character of the alloying elements. Within the AgCaLa phases, 12.5 wt% Ca, 5.1 wt% O, 4.8 wt% La, and 0.6 wt% Fe are detected via WDX. The detected O might result partially from the reaction with O of the surrounding atmosphere during or after preparation. But as the radiation evaluated during WDX measurements is generated in the surface-near area, a small amount of O is even present in the AgCaLa phases before preparation. Thus, the AgCaLa phases take up a certain amount of O during the LBM process. But a part of this O is bound in the inclusions of slag and not in the AgCaLa alloy. The La content of the AgCaLa phases proves that La almost remains in the AgCaLa alloy. A certain burning of Ca has to be assumed since Ca-rich slag is observed. Furthermore, Ca-rich slag is incorporated into the AgCaLa phases and thus, a part of the detected Ca is bound in the enclosed slag and not in the surrounding alloy. Hence, the measured value of 12.5 wt% for the AgCaLa phases is probably overestimated and Ca is lost. If necessary, the loss of Ca can be compensated by increasing the Ca content in the raw material. However, a slightly increased amount of Ca is uncritical as the electrochemical potential should be still higher than of FeMn, and a higher Ca content promotes degradability.

However, apart from the uptake of O the intended chemical composition is almost reached, as only 0.5 wt% more Ca and 0.2 wt% less La than intended are detected. Thus, the AgCaLa alloy remains as independent and adaptable phases inside the Fe matrix without any transition between AgCaLa and the matrix. As distinct diffusion of Mn is observed during the processing of FeMn with pure Ag,[68] this result is outstanding and proves the suitability of LBM to process such materials. In comparison to a conventionally processed material,[42] similar properties including a suitable electrochemical potential and degradation behavior are expectable under the assumption that the detected O originates from the preparation and the inclusions of slag and not from the uptake of O by the AgCaLa alloy itself. The hypothesis of similar properties is supported by the similarity of the structure of phases in the LBM-processed AgCaLa to the conventionally produced alloy (Figure 2d and 5a). For both materials, some compact insular-like structures, lamellae, and a third phase between the other are observed.[42] Although both structures look similar, the phases are different, as the chemical composition is different (Figure 5b,c). The lamellar structures in LBM-processed AgCaLa phases have an increased Ca content, while the lamellar

structures of the conventional alloy exhibit a reduced Ca content. Additionally, areas with high Ca and La content are detected for the additively processed material (white arrows, Figure 5b,c), whereas in the conventional alloy areas with high Ca content are separated from areas with high La content.

In summary, FeAg, as well as FeAgCaLa, exhibits satisfying properties after processing via LBM according to the requirements defined in the introduction. The AgCaLa phase remains in the intended chemical composition, but the microstructure differs from the conventionally processed alloy.[42] Thus, a secure statement regarding the properties of the AgCaLa phases is not possible; therefore, the degradation behavior is investigated in detail to clarify the properties of the AgCaLa phases. In the following section, the results of the static immersion test are presented which are influenced by the microstructure reported before.

3.4. Impact of Modification with Ag/AgCaLa on the Degradation Behavior

The temporal course of the weight changes of the samples is presented in **Figure 6**. As the differences between the three tested samples for each combination of material and electrolyte are small (average standard deviation: $3 \, g \, m^{-2}$), all courses are the average of the respective three samples. The similar corrosion behavior of the samples indicates the significance of the gained results. As all curves show an almost linear shape, the degradation rate is constant throughout the test for each material and electrolyte. Only the weight of the FeAgCaLa alloys in RL and NC increases in the first days after the start of the immersion test. Subsequently, the removal of material proceeds continuously and parallel to the degradation of the other alloys. The initially increasing weight of the FeAgCaLa samples indicates the deposition of degradation products or compounds of the immersion solutions only in conjunction with AgCaLa. However, the parallel course after the initial increase of weight indicates a similar degradation behavior of all materials investigated. Thus, the expected influence of modification with Ag and AgCaLa does not occur. At the same time, the mass loss of FeAgCaLa is the lowest in all electrolytes due to the increase in weight at the beginning.

However, removal is observed in RL and NC but not in HS. The removal in NC is slightly higher than in RL. Immersion in HS results in a continuously increasing weight, whereby the weight of FeAgCaLa is significantly more increased than the one of FeAg and Fe. The increasing weight can indicate both

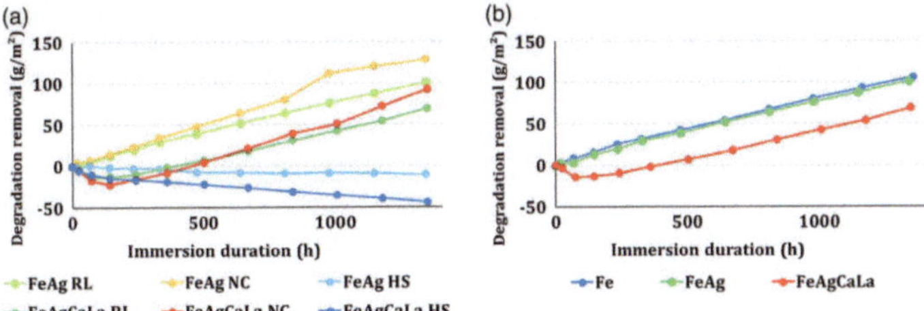

Figure 6. Temporal course of degradation removal, average of three samples for each combination: a) influence of different electrolytes on the degradation of FeAg and FeAgCaLa; b) comparison of the degradation behavior of Fe, FeAg, and FeAgCaLa in RL.

the deposition of components of the HS or degradation products. In the case of the deposition of degradation products, the increasing weight would indicate a continuously proceeding degradation. However, as the changes in weight particularly for FeAg are small, a subsequently discussed degradation inhibiting effect of some components of the HS is expectable.

As can be concluded, information concerning possibly formed depositions is crucial to explain the degradation behavior. Therefore, the samples are investigated after degradation (**Figure 7**), whereby knowledge about the local influence of the microstructure and Ag phases is gained as well. For all samples, large areas with a homogenous appearance are observed, indicating uniform degradation. However, inhomogeneity is present as well. In particular, at the edges, in the middle, or around the fixation holes of the samples, differences are apparent. Figure 7a,g) shows magnified examples of the transition between areas with different appearances. Thus, the degradation behavior differs over the sample surface.

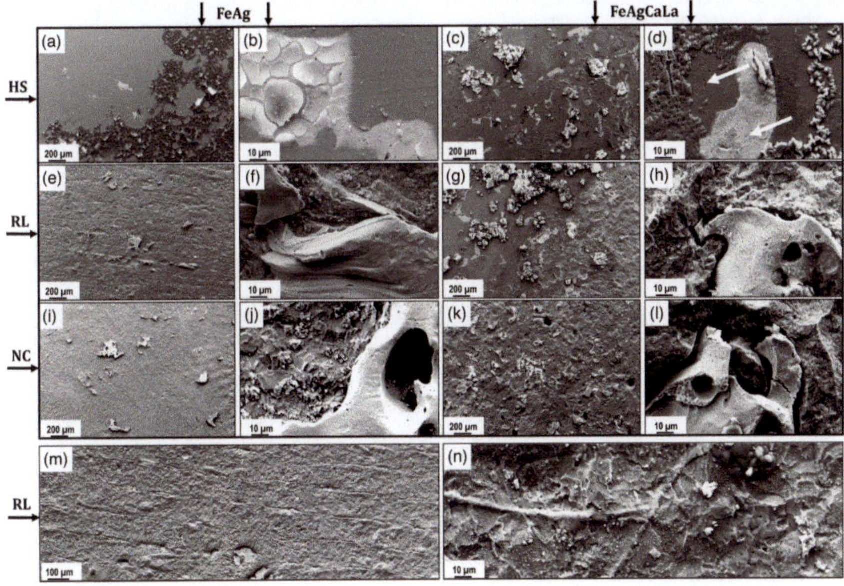

Figure 7. SEM-SE: Samples after 56 days of static immersion test: a,b) flake-like depositions; c,d) voluminous depositions, slightly corrosion; e,f) homogenous degradation apart from remaining Ag phases; g,h) areas with and without degradation apart from remaining AgCaLa phases, voluminous depositions; i,j) homogenous degradation apart from remaining Ag phases, small depositions; k,l) homogenous degradation apart from remaining AgCaLa phases, signs of the decay of AgCaLa phases; m) FeAg: visible melt tracks after immersion in RL; n) FeAg: visible grain structure after immersion.

After the immersion of FeAg in HS only, some flake-like depositions are detected following the slightly increasing weight (Figure 6a and 7a,b). Such layers are even observed on the surface of the Ag phases, indicating the formation of inhibiting layers as no signs of corrosion of the Ag phases or the matrix are detected. Thus, these layers inhibit the effect of the potential difference between Ag and Fe and the absence of increased degradation is explainable. These layers possess partially significant amounts of, e.g., Ca, P, and O which are contained in the HS. Thus, both the deposition of degradation products and components of the HS are conclusive. However, the absence of any visible sign of corrosion indicates the deposition of compounds from the HS. In contrast to FeAg, on the surface of FeAgCaLa, more, and voluminous depositions are observed which explain the increasing weight of these samples (Figure 7c,d). For FeAgCaLa, even a few signs of slight corrosion attack are visible and closed layers are not visible on the surface of the AgCaLa phases (Figure 7d). Thus, Ca and La might inhibit the formation of a layer suppressing further corrosion, whereas less detection of P on the surface of the AgCaLa phases supports this hypothesis. At the same time, there is also an increase in the formation of poorly soluble degradation products, which contain considerable amounts of Ca, P, and O. However, even for FeAgCaLa, the degradation in HS is low.

For the immersion in RL (Figure 7e–h) and NC (Figure 7i–l), a significant removal of the Fe matrix and remaining Ag and AgCaLa phases are observed. The remaining of the Ag and AgCaLa phases is proven by SEM-SE. The presented sections are representative of the appearance of the Ag and AgCaLa phases over the sample surface. But only for the FeAgCaLa alloy, a small gap between AgCaLa phases and the surrounding matrix is present, giving a hint for an increased attack of material in the direct periphery of the Ag phases. As this gap is not present before immersion (Figure 3g,h) and the AgCaLa remains unchanged, the gap is an indication of increased degradation of Fe around AgCaLa phases. However, apart from the gap formation, hints of a large-scale influence on the dissolution of Fe are not observable. Thus, the absence of any significant effect on the course of degradation (Figure 6) due to the addition of Ag and AgCaLa is confirmed.

However, as AgCaLa phases are not significantly attacked, the observations confirm the higher electrochemical potential of AgCaLa phases compared to Fe. Thus, the absence of the intended potential difference between AgCaLa phases and Fe is not a suitable explanation for the absence of the expected increased degradation. As even pure Ag phases do not significantly influence the degradation rate, the lack of potential difference as a reason for the absence of the expected effect is not conclusive anyway. Accordingly, the FeAg reference proves that the addition of Ca and La to Ag is not responsible for the absence of the expected increased degradation.

The detection of O, Ca, and a small amount of chlorine (Cl) on the surface of the Ag and AgCaLa phases after immersion in RL, as well as NC, indicates the formation of inhibiting layers like observed for HS. However, as observed for HS, these layers might be not as distinctive on the surface of AgCaLa phases as some gaps progress around the AgCaLa phases. Nevertheless, the effect of the AgCaLa phases is not strong enough to increase the degradation rate. The higher weight of

the FeAgCaLa samples is conclusive as more depositions are observed even in the areas with degradation (Figure 7e,g,i,k). Thus, the reason for the slower removal of the FeAgCaLa samples is an increased deposition which might indicate an increased degradation attack. Such an increased attack might be beneficial if the degradation products are removable in the human body.

Depositions on the surface of FeAg and FeAgCaLa contain significant amounts of Ca and O which are present in the immersion solutions. Significant amounts of La are not detected in the depositions on the surface of FeAgCaLa. Thus, the depositions are not formed due to hardly dissolvable La-containing compounds, but probably due to a changed degradation kinetic.

Subsequently, the surfaces of the samples reveal the influence of microstructure as melt tracks are observable (Figure 7m) as well as angular structures of the grains (Figure 7n).

Furthermore, hints for a beginning decay of the AgCaLa phases are present (Figure 7l) which are supported by the presence of holes (Figure 7k) eventually caused by the detachment of AgCaLa phases. The fractured surfaces visible in Figure 7l are representative for the appearance of AgCaLa phases after the immersion test. The detachment might be also due to the gaps around the AgCaLa arising from the dissolution of the Fe and not the AgCaLa phases. For reliably proof the degradation of AgCaLa phases after the matrix has dissolved, further investigations are necessary. But as the degradation rate of FeAgCaLa is not increased under the selected conditions anyway, a detailed investigation is not aimed.

Thus, the degradation of the tested material strongly depends on the electrolyte (Figure 6), and the addition of AgCaLa influences the course of degradation because increased depositions on the surface of FeAgCaLa are observed. The addition of Ag and AgCaLa has no significant effect on the degradation rate of the matrix. Probably, the formation of layers inhibits the effect of the added Ag and AgCaLa.

In summary, the results presented in this study reveal that the properties of AgCaLa phases are as intended, because a higher electrochemical potential in comparison to Fe is given. The high potential is indicated by remaining of the AgCaLa phases, whereas the signs of the beginning decay of AgCaLa phases reveal that the AgCaLa dissolves as intended after the Fe matrix. Although the loss of Ca might influence the degradability, the absence or less distinctive character of inhibiting layers disproves the remaining of AgCaLa due to protecting layers. Thus, the absence of the expected increased degradation rate is not reasoned by the deviation of properties of the FeAgCaLa from the intended properties and another hypothesis has to be assumed.

3.5. Reasoning and Suitability of Degradation Behavior

The aim of investigations under varying immersion conditions is to prove a significant, reproducible, and reliably increased degradation due to the addition of AgCaLa. As the conditions and the degradation behavior strongly differ within the human body, an increased degradation resilient to the immersion conditions is needed.[50,52,53,55,118] As described in the introduction, the need for an increased degradation is obvious and at the moment, it is sufficient to prove the accelerated degradation rate; the determination of the concrete degradation rate in the human body is not necessary.

The sensibility of the degradation behavior to the immersion conditions is underlined by the differences over the surface of samples after the immersion test. Similar effects are observed, e.g., by Schinhammer et al.[54] and Huang et al.[29] The local differences over the sample indicate different equilibria of reactions. Otherwise, a mainly homogenous removal of material without any formation of cracks or cavities is observed. Thus, the material should fulfill the requirement of homogenous degradation because no stability affecting corrosion on the local scale occurs. A later degradation of, e.g., border areas might be acceptable. However, it has to be investigated if this inhomogeneous mass removal even occurs in vivo.

Moreover, the degradation behavior differs significantly depending on the solutions, and Ca- and O-rich depositions are observed. As these depositions and the degradation products dissolved in the solution have a brown appearance, they are probably comparable to products generated in the living organism, as in vivo brown discolorations of the surrounding tissue and heavily removable degradation products are observed.[17,50,51,69] But these depositions do not inhibit further degradation because the degradation proceeds continuously. This continuous degradation is beneficial, as the load capability of the implant continually declines and the release of degradation products is continuous; thus, an oversaturation of the human organism with, e.g., Fe is not likely to occur. However, the implant is needed for a certain time, so the degradation must start in a controlled way after a defined time. So, a coating preventing degradation for a certain time might be a solution to retard the degradation at the beginning.[119]

In contrast to immersion in RL and NC, the degradation in HS is much slower. The reason for this must be the composition of the immersion solution. In contrast to RL and NC, HS contains P which reacts with Fe resulting in the formation of Fe phosphates. These phosphates are not critical as they do not suppress the exchange with the electrolyte and do not prevent further corrosion.[22] But Ca–P is formed out of Ca and P from the solution and deposited on the surface. A strong inhibiting and corrosion-preventing effect is reported for Ca–P.[15,22,49,54,56] As explained in the introduction, the addition of Ag aims to suppress the formation of such inhibiting effects, but this intended effect does not occur. This dissent is discussed in the next paragraph. However, even for in vivo experiments, in which P and Ca are present, degradation of Fe is reported. Thus, another reason must be responsible for the differing results reported in the literature. For example, the pH value significantly influences the equilibrium of chemical reactions responsible for the degradation of metallic implants, as it is decisive for the solubility of degradation products.[52,118,120] Since inflammation can cause a reduction of the pH value down to 5.5, the existence of inflammation can significantly change the degradation process.[121] On the one hand, Mandal et al.,[59] Dargusch et al.,[14] and Feng et al.[48] report an inflammatory response and degradation; on the other hand, Kraus et al.,[47] Dryndra et al.,[58] and Peuster et al.[20] could not determine any significant removal in combination with insignificant inflammation. Otherwise, Lin et al.[17] report significant degradation parallel to a slight inflammatory.

Thus, results reported in the literature are partially contradictory, and universal conditions to which implants are exposed cannot be defined. A study by Pierson et al.[50] underlines the significant influence of ambient conditions on the degradation behavior. For an identical material, degradation is observed in the arterial wall but not in the bloodstream. Thus, the significantly differing degradation, influenced by the presence of P, observed in this study is in accordance with the literature. The high degradation rate in RL and NC might be suited but is not expected to be reproducible in vivo as P is present in the human organism.

The strong influence of the immersion conditions might also explain why the increased degradation due to the addition of Ag with high electrochemical potential does not occur as reported in the literature.[14,27,28,31,32,59] The expected disturbance of inhibiting layers due to the results of Wang et al.[52] does not occur as even on the surface of the Ag phases Ca and P are detected. Similar results are reported by Loffredo et al.[122] They observed reduced degradation for material modified with Ag. Contrary to the suppression of layer formation, Wang et al.[52] guess the deposition of Ca–P on the surface of Ag might be supported by the cathodic reaction generating the necessary hydroxyl ions. After immersion in P-free solutions, a small amount of chlorine is present on the surface of the Ag phases, indicating the formation of hardly dissolvable Ag–Cl compounds which have an inhibiting effect.[123] However, less distinctive depositions are observed on the surface of the AgCaLa phases. Thus, inhibiting layers on the surface of AgCaLa are less pronounced than on the surface of Ag. As an inhibiting effect is even present for the FeAgCaLa, another effect might be responsible for the absence of the expected anodic dissolution of Fe. Inhibiting layers on the surface of Fe are a possible explanation. To confirm this assumption further, detailed investigations of the surface after the immersion test are necessary.

However, the development of gaps around AgCaLa phases is a hint of an effect of the different potentials. But this effect is not strong enough to increase the degradation rate significantly. Nevertheless, the degradation of FeAgCaLa might be increased without a noticeable effect on the weight loss due to increased depositions. Dargusch et al.[14] reported gaps around the Ag phases as observed in this study after immersion test and increased degradation rate. But it has to be considered that the Ag phases observed by Dargusch et al.[14] are smaller, and thus, the impact of the gaps might be higher. Accordingly, the phases obtained via LBM might be too large and have to be adjusted by a parameter adaption or another manufacturing technique has to be applied.

In summary, the concept of Fe modified with Ag to adapt the degradation failed under the conditions selected for the immersion test. But that does not mean the complete failure of the strategy, as the immersion conditions strongly affect the material removal, and thus, acceleration due to the addition of Ag might be possible under some conditions in the human organism as reported in the literature.[14,27,28,31,32] Thus, the material might be suited for some applications. However, the concept does not fulfill the required reliable increased mass removal and thus, an adaption of the material is appropriate. As even the microstructure of pure Fe without any segregation is visible, the structure of the matrix influences the degradation, and the adaption of the matrix is aimed. Therefore, alloying with Mn seems to be a promising strategy as stated in the introduction and the adaption of the matrix might be more promising than a further investigation of pure Fe "alloyed" with Ag.[14–17,21–23] Thus, the findings of the present study can be utilized to further adapt materials for Fe-based degradable material or can be transferred to other applications of immiscible material combinations.

4. Conclusions

The presented investigations aim to achieve degradable AgCaLa phases within a Fe matrix via LBM and examine the effect on the degradation behavior. The processing of the mixed material via LBM is successfully performed, whereas the expected increased dissolution of the Fe matrix does not occur. The findings of this study can be summarized as follows: 1) Processing of pure Fe with degradable AgCaLa alloy via LBM is possible. The AgCaLa remains as an adaptable phase with intended morphology after the LBM process. 2) The intended chemical composition of AgCaLa is almost achieved. Although the AgCaLa phases are impaired by inclusions of slag, they still have a high electrochemical potential and degradability can be assumed. 3) The morphology and mechanism of the formation of Ag phases significantly differ from FeAg to FeAgCaLa due to vapor pressure and surface tension-driven effects. The AgCaLa phases show a distinct similarity to Ag phases in FeMn.[68] 4) The immersion conditions significantly influence the degradation behavior. If Ca and P, as ingredients of the HS, are present, corrosion inhibiting layers of Ca–P compounds are observed even on the surface of the Ag and AgCaLa phases. 5) The modification of Fe with Ag does not increase the degradation rate under the selected immersion conditions. The locally different electrochemical potential does not result in the anodic dissolution of the Fe matrix. 6) On the surface of AgCaLa phases, less significant depositions are formed and small gaps around the AgCaLa phases develop indicating an effect of the potential difference. But due to increased depositions on the surface of FeAgCaLa, an increased mass loss could not be observed.

Thus, the results highlight the suitability of LBM to process immiscible material combinations like Fe with an AgCaLa alloy as the Ag alloy remains as an adaptable phase, and no balance of alloying elements by diffusion occurs. Besides, the results of the immersion test illustrate the need for further investigations and adaption of the material, e.g., by alloying the matrix with Mn.

Acknowledgements

The authors gratefully thank the German Research Foundation (DFG) for financial support. This work results from the project SCHA1484/44-1. Furthermore, the authors would like to thank Jessica Heinrich, Eva Mühlhoff, Lea Kaspersmeier, and Anja Puda for promoting the production and examination of the tested material.

Open Access funding enabled and organized by Projekt DEAL.

Conflict of Interest

The authors declare no conflict of interest.

Data Availability Statement

Research data are not shared.

Keywords

biomedical application, bioresorbable metal, corrosion, iron alloys, laser beam melting, silver alloys

Received: July 12, 2022
Revised: September 7, 2022
Published online:

[1] S. Veerachamy, T. Yarlagadda, G. Manivasagam, P. K. D. V. Yarlagadda, *Proc. Inst. Mech. Eng. H* **2014**, *228*, 1083.

[2] Q. Chen, G. A. Thouas, *Mater. Sci. Eng. R: Rep.* **2015**, *87*, 1.

[3] C. Mauffrey, B. Herbert, H. Young, M. L. Wilson, M. Hake, P. F. Stahel, *Eur. J. Trauma Emerg. Surg.* **2016**, *42*, 411.

[4] F. Zivic, S. Affatato, M. Trajanovic, M. Schnabelrauch, N. Grujovic, K. L. Choy, *Biomaterials in Clinical Practice*, Springer, Cham **2018**.

[5] D. Carluccio, A.G. Demir, M. J. Bermingham, M. S. Dargusch, *Metall. Mater. Trans. A* **2020**, 3311.

[6] D. R. Garcia, D. G. Deckey, A. Zega, C. Mayfield, C. S. L. Spake, T. Emanuel, A. Daniels, J. Jarrell, J. Glasser, C. T. Born, C. P. Eberso, *Spine Deform.* **2020**, *8*, 351.

[7] H. Hermawan, *Prog. Biomater.* **2018**, *7*, 93.

[8] M. Prakasam, J. Locs, K. Salma-Ancane, D. Loca, A. Largeteau, L. Berzina-Cimdina, *J. Funct. Biomater.* **2017**, *8*, 44.

[9] M. Alonzo, F. A. Primo, S. A. Kumar, J. A. Mudloff, E. Dominguez, G. Fregoso, N. Ortiz, W. M. Weiss, B. Joddar, *Curr. Opin. Biomed. Eng.* **2021**, *17*, 100248.

[10] E. Zhang, X. Zhao, J. Hu, R. Wang, S. Fu, G. Qin, *Bioact. Mater.* **2021**, *6*, 2569.

[11] L. Pleva, P. Kukla, O. Hlinomaz, *J. Geriatr. Cardiol.* **2018**, *15*, 173.

[12] Y. F. Zheng, X. N. Gu, F. Witte, *Mater. Sci. Eng. R: Rep.* **2014**, *77*, 1.

[13] H. Hermawan, D. Dubé, D. Mantovani, *Adv. Mater. Res.* **2007**, *15–17*, 107.

[14] M. S. Dargusch, J. Venezuela, A. Dehghan-Manshadi, S. Johnston, N. Yang, K. Mardon, C. Lau, R. Allavena, *Adv. Healthcare Mater.* **2020**, *10*, 202000667.

[15] S. Mandal, Viraj, S. K. Nandi, M. Roy, *J. Mater. Chem. C* **2021**, *9*, 4340.

[16] N. Abbaspour, R. Hurrell, R. Kelishadi, *J. Res. Med. Sci.* **2014**, *19*, 164.

[17] W. Lin, L. Qin, H. Qi, D. Zhang, G. Zhang, R. Gao, H. Qiu, Y. Xia, P. Cao, X. Wang, W. Zheng, *Acta Biomater.* **2017**, *54*, 454.

[18] P. Trumbo, S. Schlicker, A. A. Yates, M. Poos, *J. Am. Diet. Assoc.* **2002**, *102*, 1621.

[19] G. Poologasundarampillai, A. Nommeots-Nomm, *3D Print. Med.* **2017**, 43.

[20] M. Peuster, C. Hesse, T. Schloo, C. Fink, P. Beerbaum, C. von Schnakenburg, *Biomater.* **2006**, *26*, 4955.

[21] M. S. Dargusch, A. Dehghan-Manshadi, M. Shahbazi, J. Venezuela, X. Tran, J. Song, N. Liu, C. Xu, Q. Ye, C. Wen, *ACS Biomater. Sci. Eng.* **2019**, 1686.

[22] C. Tonna, C. Wang, D. Mei, S. V. Lamaka, M. L. Zheludkevich, J. Buhagiar, *Bioact. Mater.* **2022**, *7*, 426.

[23] M. Schinhammer, A. C. Hänzi, J. F. Löffler, P. J. Uggowitzer, *Acta Biomater.* **2010**, *6*, 1705.

[24] M. Schinhammer, I. Gerber, A. C. Hänzi, P. J. Uggowitzer, *Mater. Sci. Eng. C* **2013**, *33*, 782.

[25] J. Crossgrove, W. Zheng, *NMR Biomed.* **2004**, *17*, 544.

[26] J. Cheng, B. Liu, Y. H. Wu, Y. F. Zheng, *J. Mater. Sci. Technol.* **2013**, *29*, 619.

[27] T. Niendorf, F. Brenne, P. Hoyer, D. Schwarze, M. Schaper, R. Grothe, M. Wiesener, G. Grundmeier, H. J. Maier, *Metall. Mater. Trans. A* **2015**, *46*, 2829.

[28] T. Huang, J. Cheng, D. Bian, Y. Zheng, *J. Biomed. Mater. Res.* **2015**, *104*, 225.

[29] T. Huang, J. Cheng, Y. F. Zheng, *Mater. Sci. Eng. C* **2014**, *35*, 43.

[30] J. Čapek, Š. Msallamová, E. Jablonská, J. Lipovc, D. Vojtěch, *Mater. Sci. Eng. C* **2017**, *79*, 550.

[31] J. He, F. L. He, D. W. Li, Y. L. Liu, Y. Y. Liu, Y. J. Ye, D. C. Yin, *RSC Adv.* **2016**, *6*, 112819.

[32] P. S. Bagha, M. Khakbiz, S. Sheibani, H. Hermawan, *J. Alloys Compd.* **2018**, *767*, 955.

[33] R. Y. Liu, R. G. He, Y. X. Chen, S. F. Guo, *Acta Metall. Sin.* **2019**, *32*, 1337.

[34] K Mijnendonckx, N. Leys, J. Mahillon, S. Silver, R. V. Houdt, *Biometals* **2013**, *26*, 609.

[35] W. C. Chiang, I-S. Tseng, P. Møller, L. Rischel Hilbert, T. T. Nielsen, J. K. Wu, *Mater. Chem. Phys.* **2010**, *119*, 123.

[36] World Health Organization, *Silver in Drinking-Water Background Document for Development of WHO Guidelines for Drinking-Water Quality Guidelines for Drinking-Water Quality, Second edn., Health Criteria and Other Supporting Information 2*, World Health Organization, Geneva **1996**.

[37] L. Bonilla Gameros, P. Chevallier, A. Sarkissian, D. Mantovani, *Biol. Med.* **2020**, *24*, 102142.

[38] D. J. Hall, R. Pourzal, J. J. Jacobs, R. M. Urban, *J. Biomed. Mater. Res. B* **2019**, *107*, 1930.

[39] R. M. Urban, J. J. Jacobs, M. J. Tomlinson, J. Gavrilovic, J. Black, M. Peoc'h, *J. Bone Joint Surg. A* **2000**, *82*, 457.

[40] M. Heiden, E. Walker, L. Stanciu, *J. Biotechnol. Biomater.* **2015**, *05*, 1000178.

[41] M. Afifi, S. Saddick, O. A. Zinada, *Saudi J. Biol. Sci.* **2016**, *23*, 754.

[42] J. T. Krüger, K. P. Hoyer, V. Filor, S. Pramanik, M. Kietzmann, J. Meißner, M. Schaper, *J. Alloys Compd.* **2021**, *871*, 159544.

[43] A. Andreiev, K. P. Hoyer, O. Grydin, Y. Frolov, M. Schaper, *Mater. Werkstofftech.* **2020**, *51*, 517.

[44] J. T. Krüger, K. P. Hoyer, M. Schaper, *Mater. Lett.* **2022**, *306*, 130890.

[45] K. P. Hoyer, M. Schaper, in: *TMS 2019 148th Annual Meeting & Exhibition Supplemental Proceedings*. The Minerals, Metals & Materials Series. Springer, Cham **2019**.

[46] M. Peuster, P. Wohlsein, M. Brügmann, M. Ehlerding, K. Seidler, C. Fink, H. Brauer, A. Fischer, G. Hausdorf, *Heart* **2001**, *86*, 563.

[47] T. Kraus, F. Moszner, S. Fischerauer, M. Fiedler, E. Martinelli, J. Eichler, F. Witte, E. Willbold, M. Schinhammer, M. Meischel, P. J. Uggowitzer, J. F. Löffler, A. Weinberg, *Acta Biomater.* **2014**, *10*, 3346.

[48] Q. Feng, D. Zhang, C. Xin, X. Liu, W. Lin, W. Zhang, S. Chen, K. Sun, *J. Mater. Sci.: Mater. Med.* **2013**, *24*, 713.

[49] B. Paul, A. Lode, A. M. Placht, A. Voß, S. Pilz, U. Wolff, S. Oswald, A. Gebert, M. Gelinsky, J. Hufenbach, *ACS Appl. Mater. Interfaces* **2021**, *14*, 439.

[50] D. Pierson, J. Edick, A. Tauscher, E. Pokorney, P. Bowen, J. Gelbaugh, J. Stinson, H. Getty, C. H. Lee, J. Drelich, J. Goldman, *J. Biomed. Mater. Res., B* **2011**, 58.

[51] R. Waksman, R. Pakala, R. Baffour, R. Seabron, D. Hellinga, F. O. Tio, *J. Interv. Cardiol.* **2007**, *21*, 15.

[52] C. Wang, C. Tonna, D. Mei, J. Buhagiar, M. L. Zheludkevich, S. V. Lamaka, *Bioact. Mater.* **2022**, *7*, 412.

[53] J. Huang, A. G. Orive, J. T. Krüger, K. P. Hoyer, A. Keller, G. Grundmeier, *Corros. Sci.* **2022**, *200*, 110186.

[54] M. Schinhammer, P. Steiger, F. Moszner, J. F. Löffler, P. J. Uggowitzer, *Mater. Sci. Eng. C* **2013**, *33*, 1882.

[55] P. S. Bagha, M. Khakbiz, S. Sheibani, S. Ebrahimi-Barough, H. Hermawan, *ACS Biomater. Sci. Eng.* **2020**, 2094.

[56] H. Hermawan, A. Purnama, D. Dube, J. Couet, D. Mantovani, *Acta Biomater.* **2010**, *6*, 1852.

[57] J. Hufenbach, J. Sander, F. Kochta, S. Pilz, A. Voss, U. Kühn, A. Gebert, *Adv. Eng. Mater.* **2022**, *22*, 2000182.

[58] A. Drynda, T. Hassel, F. W. Bach, M. Peuster, *J. Biomed. Mater. Res., Part B.* **2014**, *103*, 649.

[59] S. Mandal, V. Kishore, M. Bose, S. K. Nandi, M. Roy, *J. Mater. Sci. Technol.* **2021**, *84*, 159.

[60] J. Z. Zhao, T. Ahmed, H. X. Jiang, J. He, Q. Sun, *Acta Metall. Sin.* **2017**, *30*, 1.

[61] ASM Alloy P D D ASM Alloy Phase Diagram Database, **1999**.

[62] C. Wei, J. Wang, Y. He, J. Li, E. Beaugnon, *Metals* **2021**, *11*, 525.

[63] M. C Conti, B. Mallia, E. Sinagra, P. Schembri Wismayer, J. Buhagiar, D. Vella, *Heliyon* **2019**, *5*, 02522.

[64] P. Sotoudehbagha, S. Sheibanib, M. Khakbiz, S. Ebrahimi-Barough, H. Hermawan, *Mater. Sci. Eng., C.* **2018**, *88*, 88.

[65] J. Kraner, J. Medved, M. Godec, I. Paulin, *Metals* **2021**, *11*, 234.

[66] J. Quan, K. Lina, D. Gua, *Powder Technol.* **2020**, *364* 478.

[67] S. Dadbakhsh, R. Mertens, L. Hao, J. Van Humbeeck, J. P. Kruth, *Adv. Eng. Mater.* **2019**, *21*, 1801244.

[68] J. T. Krüger, K. P. Hoyer, F. Hengsbach, M. Schaper, *J. Mater. Res. Technol.* **2022**, *19C*, 2369.

[69] X. Guo, C. Zhang, Q. Tian, D. Yu, *Mater. Today Commun.* **2021**, *26*, 102007.

[70] I. McCue, B. Gaskey, P. A. Geslin, A. Karma, J. Erlebacher, *Acta Mater.* **2016**, *115*, 10.

[71] G. Mikhailov, L. Makrovets, *Russ. Metall.* **2008**, *2008*, 727.

[72] R. J. Fruehan, *Metall. Trans.* **1974**, *5*, 345.

[73] Z. Ferdous, A. Nemmar, *Int. J. Mol. Sci.* **2020**, *21*, 2375.

[74] D. Wang, S. Wu, F. Fu, S. Mai, Y. Yang, Y. Liu, C. Song, *Mater. Des.* **2017**, *117*, 121.

[75] S. A. Khairallah, A. T. Anderson, A. Rubenchik, W. E. King, *Acta Mater.* **2016**, *108*, 36.

[76] S. A. Khairallah, A. Anderson, *J. Mater. Process. Technol.* **2014**, *214*, 2627.

[77] X. Zhou, K. Li, D. Zhang, X. Liu, J. Ma, W. Liu, Z. Shen, *J. Alloys Compd.* **2015**, *631*, 153.

[78] E. L. Papazoglou, N. E. Karkalos, P. Karmiris-Obratański, A. P. Markopoulos, *Arch. Comput. Meth. Eng.* **2022**, *29*, 941.

[79] V. Gunenthiram, P. Peyre, M. Schneider, M. Dal, F. Coste, I. Koutiri, R. Fabbro, *J. Mater. Process. Technol.* **2017**, *251*, 376.

[80] M. J. Matthews, G. Guss, S. A. Khairallah, A. M. Rubenchik, P. J. Depond, W. E. King, *Acta Mater.* **2016**, *114*, 33.

[81] T. Heeling, M. Cloots, K. Wegener, *Addit. Manuf.* **2017**, *14*, 116.

[82] V. Gunenthiram, P. Peyre, M. Schneider, M. Dal, F. Coste, R. Fabbro, *J. Laser Appl.* **2017**, *29*, 022303.

[83] A. Bin Anwar, Q. C. Pham, *J. Mater. Process. Technol.* **2017**, *240*, 388.

[84] S. Shrestha, Y. K. Chou, *J. Manuf. Sci. Eng.* **2019**, *141*, 101002.

[85] L. Huanga, X. Hua, D. Wu, L. Fang, Y. Cai, Y. Ye, *J. Manuf. Process.* **2018**, *33* 43.

[86] A. Queva, G. Guillemot, C. Moriconi, C. Metton, M. Bellet, *Addit. Manuf.* **2020**, *35*, 101249.

[87] M. Miyagi, H. Wang, R. Yoshida, Y. Kawahito, H. Kawakami, T. Shoubu, *Sci. Rep.* **2018**, *8*, 12944.

[88] J. Trapp, A. M. Rubenchik, G. Guss, M. J. Matthews, *Appl. Mater.* **2017**, *9*, 341.

[89] W. E. King, H. D. Barth, V. M. Castillo, G. F. Gallegos, J. W. Gibbs, D. E. Hahn, C. Kamath, A. M. Rubenchik, *J. Mater. Process. Technol.* **2014**, *214*, 2915.

[90] S. Patel, M. Vlasea, *Mater.* **2020**, *9*, 100591.

[91] C. Qiu, C. Panwisawas, M. Ward, H. C. Basoalto, J. W. Brooks, M. M. Attallah, *Acta Mater.* **2015**, *96*, 72.

[92] M. Lutter-Günther, M. Bröker, T. Mayer, S. Lizak, C. Seidel, G. Reinhart, *Proc. CIRP* **2018**, *74*, 33.

[93] M. T. Andani, R. Dehghani, M. R. Karamooz-Ravari, R. Mirzaeifar, J. Ni, *Addit. Manuf.* **2018**, *20*, 33.

[94] J. Volpp, *J. Laser Appl.* **2020**, *32*, 022023.

[95] E. Louvis, P. Fox, C. J. Sutcliffe, *J. Mater. Process. Technol.* **2011**, *211*, 275.

[96] H. Yu, S. Hayashi, K. Kakehi, Y. L. Kuo, *Metals* **2019**, *9*, 19.

[97] M. Salehi, S. Maleksaeedi, H. Farnoush, N. M. L. Sharon, G. K. Meenashisundaram, M. Gupta, *Powder Technol.* **2018**, *333*, 252.

[98] A. Ghasemi, E. Fereiduni, M. Balbaa, S. D. Jadhav, M. Elbestawi, S. Habibi, *Addit. Manuf.* **2021**, *46*, 102145.

[99] W. Zhang, L. Wang, Z. Feng, Y. Chen, *Optik* **2020**, *207*, 163842.

[100] L. Cao, S. Chen, M. I. Wei, Q. Guo, J. Liang, C. Liu, M. Wang, *Opt. Laser Technol.* **2019**, *111*, 541.

[101] X. Shi, C. Yan, W. Feng, Y. Zhang, Z. Leng, *Opt. Laser Technol.* **2020**, *132*, 106471.

[102] G. Shannon, L. White, S. Sridhar, *J. Mater. Sci. Eng. A* **2008**, *495*, 310.

[103] O. Andreau, I. Koutiri, P. Peyre, J. D. Penot, N. Saintier, E. Pessard, T. De Terris, C. Dupuy, T. Baudin, *J. Mater. Process. Technol.* **2019**, *264*, 21.

[104] P. Lejček, M. Roudnická, J. Čapek, D. Dvorský, J. Drahokoupil, D. Šimek, J. Čížek, P. Svora, O. Molnárová, D. Vojtěch, *Mater. Charact.* **2019**, *154*, 222.

[105] I. Barin, O. Knacke, in *Thermochemical Properties of Inorganic Substances*, Springer-Verlag, Berlin, **1973**.

[106] I. Egry, *Int. J. Thermophys.* **2005**, *26*, 931.

[107] J. Srinivasan, B. Basu, *Int. J. Heat Mass Transfer* **1986**, *29*, 563.

[108] S. Ozawa, K. Morohoshi, T. Hibiya, H. Fukuyama, *J. Appl. Phys.* **2010**, *107*, 014910.

[109] S. Ozawa, S. Takahashi, S. Suzuki, H. Sugawara, H. Fukuyama, *Jpn. J. Appl. Phys.* **2011**, *50*, 11RD05.

[110] I. Egry, E. Ricci, R. Novakovic, S. Ozawa, *Adv. Colloid Interface Sci.* **2010**, *159*, 198.

[111] T. Tanaka, K. Hack, S. Hara, *MRS Bulletin* **1999**, *24*, 45.

[112] H. M. Lu, Q. Jiang, *J. Phys. Chem. B* **2005**, *109* 15463.

[113] T. Hibiya, S. Ozawa, *Cryst. Res. Technol.* **2013**, *48*, 208.

[114] R. U. Ahsan, M. Cheepu, R. Ashiri, T. H. Kim, C. Jeong, Y. D. Park, *Weld. World* **2017**, *61*, 1275.

[115] D. W. Cho, Y. D. Park, M. Cheepu, *Int. Commun. Heat Mass Transfer* **2021**, *125*, 105243.

[116] A. Dass, A. Moridi, *Coatings* **2019**, *9*, 418.

[117] Z. S. Saldi, A. Kidess, S. Kenjereš, C. Zhao, I. M. Richardson, C. R. Kleijn, *Int. Commun. Heat Mass Transfer* **2013**, *66*, 879.

[118] M. Pourbaix, in *Aqueous Solutions, Second English Edition* National Association of Corrosion Engineers, Houston, **1974**.

[119] M. Moravej, D. Mantovani, *Int. J. Mol. Sci.* **2011**, *12*, 4250.

[120] V. S. Saji, C. W. Lee, *ChemSusChem* **2012**, *5*, 1146.

[121] K. E. Barrett, S. M. Barman, J. Yuan, H. L. Brooks, in *Ganong's Review of Medical Physiology* McGraw Hill/Medical, New York, NY **2019**.

[122] S. Loffredo, C. Paternoster, N. Giguère, M. Vedani, D. Mantovani, *JOM* **2020**, *72*, 1892.

[123] J. F. Liebman, *Struct. Chem.* **2004**, *15*, 165.

Journal of
*Functional
Biomaterials*

Article

FeMn with Phases of a Degradable Ag Alloy for Residue-Free and Adapted Bioresorbability

Jan Tobias Krüger [1,2,*], Kay-Peter Hoyer [1,2], Jingyuan Huang [3], Viviane Filor [4], Rafael Hernan Mateus-Vargas [5], Hilke Oltmanns [5], Jessica Meißner [5], Guido Grundmeier [3] and Mirko Schaper [1,2]

1   Materials Science, Paderborn University, Mersinweg 7, 33100 Paderborn, Germany
2   DMRC-Direct Manufacturing Research Center, Paderborn University, Mersinweg 3, 33100 Paderborn, Germany
3   Technical and Macromolecular Chemistry, Paderborn University, Warburger Str. 100, 33098 Paderborn, Germany
4   Department of Veterinary Medicine, Institute of Pharmacology and Toxicology, Freie Universität Berlin, Koserstr. 20, 14195 Berlin, Germany
5   Department of Pharmacology, Toxicology and Pharmacy, University of Veterinary Medicine Hannover, Foundation, Buenteweg 17, 30559 Hannover, Germany
*   Correspondence: krueger@lwk.upb.de

Abstract: The development of bioresorbable materials for temporary implantation enables progress in medical technology. Iron (Fe)-based degradable materials are biocompatible and exhibit good mechanical properties, but their degradation rate is low. Aside from alloying with Manganese (Mn), the creation of phases with high electrochemical potential such as silver (Ag) phases to cause the anodic dissolution of FeMn is promising. However, to enable residue-free dissolution, the Ag needs to be modified. This concern is addressed, as FeMn modified with a degradable Ag-Calcium-Lanthanum (AgCaLa) alloy is investigated. The electrochemical properties and the degradation behavior are determined via a static immersion test. The local differences in electrochemical potential increase the degradation rate (low pH values), and the formation of gaps around the Ag phases (neutral pH values) demonstrates the benefit of the strategy. Nevertheless, the formation of corrosion-inhibiting layers avoids an increased degradation rate under a neutral pH value. The complete bioresorption of the material is possible since the phases of the degradable AgCaLa alloy dissolve after the FeMn matrix. Cell viability tests reveal biocompatibility, and the antibacterial activity of the degradation supernatant is observed. Thus, FeMn modified with degradable AgCaLa phases is promising as a bioresorbable material if corrosion-inhibiting layers can be diminished.

Keywords: antibacterial behavior; biocompatibility; biomedical application; bioresorbable metal; corrosion; iron alloys; silver alloys

Citation: Krüger, J.T.; Hoyer, K.-P.; Huang, J.; Filor, V.; Mateus-Vargas, R.H.; Oltmanns, H.; Meißner, J.; Grundmeier, G.; Schaper, M. FeMn with Phases of a Degradable Ag Alloy for Residue-Free and Adapted Bioresorbability. *J. Funct. Biomater.* **2022**, *13*, 185. https://doi.org/10.3390/jfb13040185

Academic Editor: Lidy Fratila-Apachitei

Received: 30 August 2022
Accepted: 5 October 2022
Published: 13 October 2022

Publisher's Note: MDPI stays neutral with regard to jurisdictional claims in published maps and institutional affiliations.

1. Introduction

Evermore diseases and injuries can be treated with implants to improve the quality of a patient's life. Therefore, adapted materials to produce implants with suited properties for each application are necessary [1–4]. Most of the materials are developed by focusing on inert material behavior to guarantee biocompatibility, without considering the further potentially beneficial properties of materials [1,2,5,6]. Thus, the focus of actual research is to create implants with adapted stiffness, antibacterial properties, or the development of materials that enable bioresorption [3,7–9]. Bioresorbable implants degrade, while the remaining degradation products can be removed and excreted by the human body [1,4,7]. Accordingly, they are beneficial for applications where implants are only needed for a certain time such as a fixture of bone fracture in osteosynthesis. If these implants remain longer than necessary in the human body, they can lead to adverse consequences, and the removal is connected to patient burden and operational risk [3,7–9]. Some degradable materials, such as magnesium (Mg) alloys or polymers, are already used successfully. Nevertheless, they do not fulfill the requirements for all applications, such as the treatment of bone fracture or arteriosclerosis with stents. Thus, further materials such as molybdenum (Mo), zinc (Zn), and iron (Fe) are investigated [1,2,4,7,10–12]. Due to its mechanical properties and biocompatibility, Fe is a promising material [10,13–16]. Since Fe is a necessary trace element for the human organism, with a needed uptake of 8–27 mg per day corresponding to 9.8 g per year, the amount of Fe released by the dissolution of an osteosynthesis plate with a weight of 2 g over a few months is acceptable [17].

However, an adaption of pure Fe is necessary, since the degradation rate is too low, the mechanical properties are insufficient compared to commercially used Fe alloys, and the use of magnetic resonance imaging (MRI) is impossible. [3,10,12,18] Alloying with manganese (Mn) is a promising approach to overcome these limitations [13,14,19,20]. Mn stabilizes the paramagnetic austenitic structure, enabling MRI and improving the mechanical properties. High amounts of Mn evoke the effect of twinning-induced plasticity (TWIP), resulting in good strength and formability [12,21–23]. Moreover, a reduction in the electrochemical potential is obtained by the addition of Mn, causing an increased degradation rate compared to pure Fe [12,19,24]. Segregations of Mn result in local differences in the electrochemical potential and subsequently enhance the anodic dissolution, resulting in an overall increased degradation rate [12,16,20,24,25]. Shuai et al. [25], for instance, observed an approximately 2.5-times-higher degradation rate for Fe alloyed with 25 wt.-% Mn compared to pure Fe after a one-month static immersion test. However, the biocompatibility of Mn is critically discussed, and a dose of 4 mg per day corresponding to 1.5 g per year should not be exceeded [26,27]. The addition of carbon (C) reduces the Mn-amount necessary to achieve the TWIP-effect, causing enhanced biocompatibility [22,23]. Thus, steel containing Mn and C showing the TWIP-effect is selected for the present study.

Since, during degradation, the released ions react and form corrosion products, the consideration of the biocompatibility of single elements is not sufficient, and corrosion products generated in the complex environment of the living organism must be considered. Investigations of biocompatibility, even in vivo, consider the good biocompatibility of FeMn [13,19,27–31].

To fulfill the aim of bioresorbability, the degradation products must be excreted completely. Hardly dissolvable degradation products are observed at the implantation site of Fe-based stents, resulting in the brown discoloration of the tissue [29,32–34]. Since degradation products transported to the lymph nodes are observed 52 months after the implantation of Fe-stents, excretion by the lymphatic system is expectable. As a decomposition by macrophages is expectable as well, complete bioresorption should occur [16,32,35].

However, Dryndra et al. [28] reported biocompatibility but no significant degradation for in vivo-tested FeMn. For most applications, the degradation rate of FeMn is still too low, and an increase in the degradation rate is necessary. For the adaption of the degradation rate, the concentration of released ions during degradation has to be considered, as it is a decisive factor for biocompatibility [16,19,23,28,31]. The creation of phases with different electrochemical potentials to cause the anodic dissolution of less noble phases can increase the overall degradation rate. Thus, the addition of phases with high electrochemical potential to an FeMn matrix is aimed for. This strategy enables increased degradation, while the FeMn matrix with well-suited properties remains unchanged [22,36–38]. Noble elements with high electrochemical potential are promising in forming such phases. Due to the immiscibility of silver (Ag) with Fe in the liquid and solid states, Ag phases can exist inside an unchanged FeMn matrix. Additionally, Ag is biocompatible and has antibacterial properties [31,39–41]. Reports in the literature confirm an increased degradation rate due to the modification with Ag [22,31,35,39,42]. Dargush et al. [35] report an almost twice-as-large degradation rate after 12 weeks of in vivo degradation of Fe alloyed with 35 wt.-% Mn containing 1 wt.-% Ag compared to material without Ag. According to Huang et al. [39]

the degradation rate of Fe with 5 wt.-% Ag is 50% higher than that for pure Fe after 30 days of immersion in Hank's solution.

However, further aspects have to be considered to design such material. The various elements and chemical compounds in the human body result in the formation of various corrosion products, forming a multilayer structure [36,43,44]. These structures might affect further degradation. The released ions can form phosphates which deposit on the surface. Compounds with Fe and Mn are present on the surface, but the degradation can proceed further since exchange with the electrolyte is still possible [16,19,30,44]. In contrast, Ca-Phosphate (P), originating from the electrolyte, creates an inhibiting layer deteriorating the degradation rate [13,19,28,30,36,45]. Therefore, the formation of such inhibiting layers must be suppressed to enable further degradation. Wang et al. [44] suggest that noble phases can minimize the blocking effect, as a detachment of inhibiting layers and a changed structure can be observed. This suggestion is confirmed by the increased degradation rate (in vivo) of materials modified with Ag [35,38].

However, pure Ag is not suitable for modifying FeMn since the Ag particles would remain after the degradation of the matrix. This is not acceptable, as impairments may occur due to the remaining particles, and this contradicts the avoidance of such risks by complete bioresorption [46]. As particles can be transported through the human body, Ag might reach areas such as the brain, where the release of Ag is critically discussed [47–49]. Furthermore, impairments due to tissue damage by sharp edges, thrombosis, or strokes are possible [50,51]. The replacement of pure Ag with a completely degradable Ag alloy is an approach to prevent such consequences. Such alloy enables the degradation of Ag phases after the matrix, while their electrochemical potential is high enough to enhance the anodic dissolution of the FeMn matrix [41,52,53]. To the best of the authors' knowledge, this study is the first to report on the modification of FeMn with a degradable Ag alloy, enabling a residue-free dissolution. Thus, the concern of remaining Ag particles after the degradation of the Fe-based matrix is addressed.

For the adaption of degradable Ag phases, it has to be considered that the biocompatibility of Ag depends on the concentration of released Ag ions and the total released amount of Ag [40,54,55]. If the concentration of Ag is too high, the mechanism responsible for the antibacterial effect harms the human cell, but a suitable concentration of Ag enables antibacterial properties without impairment of the organism [56–58]. Since implant-related infections are a major challenge for the use of implants, the simultaneous addressing of this issue is one major benefit of the addition of Ag [1,58,59]. The adjustment of the Ag release through alloying is therefore purposeful. A degradable Ag alloy containing calcium (Ca) and lanthanum (La) (AgCaLa) developed by Krüger et al. [41] is proven to be suitable for modifying FeMn. Ca is a necessary trace element, and La is successfully investigated as an alloying element of degradable Mg alloys, although the uptake should be strictly limited [60,61].

The fabrication of immiscible material combinations such as Fe and Ag is challenging. Powder-based processes such as the additive manufacturing method laser beam melting (LBM) enable the processing of such materials since powders can be mixed mechanically. During the subsequent continuous melting process characterized by a small melt pool, a strong melt flow, and rapid melting and solidification, the Ag is incorporated into the bulk material. By the modification of the raw material and the LBM process, the chemical composition and the morphology of the Ag and AgCaLa phases are adaptable [22,62–64]. The raw material for the AgCaLa phases is obtained by conventional alloying and subsequent gas atomization, whereas the final structure and chemical composition of the AgCaLa inside the FeMn matrix are influenced by the LBM process [41,64].

For the specific adjustment of the degradation behavior, the varying conditions in the human organism and the individual response of the human organism have to be considered. Achieving a reliable and suited degradation under all possible conditions is challenging, since the degradation of Fe-based alloys strongly depends on the ambient conditions. The degradation is affected by the ingredients of the solution, the pH value, which determines the solubility of the corrosion products, and the amount of dissolved gases, such as oxygen (O), that is required for the cathodic reaction [14,33,35,37,44]. For implant design, the slowest degradation must guarantee a sufficient supporting effect for healing, while the fastest degradation must enable the dissolution in a defined period.

However, the increased degradation of the FeMn matrix, the degradation of the Ag-CaLa particles after the matrix, biocompatibility, and an antibacterial effect are expectable but not guaranteed for FeMn modified with degradable AgCaLa. Therefore, an in-depth investigation of the listed properties is necessary, and the corresponding results are presented in this study. These results enable the further targeted adaption of the material to fulfill all requirements.

2. Materials and Methods

The investigated samples are produced via LBM using a mixture of FeMn (nominal composition: 22 wt.-% Mn, 0.6 wt.-% C and Fe bal.; Nanoval GmbH & Co. KG, Berlin, Germany) powder with 5 wt.-% AgCaLa (nominal composition: 12 wt.-% Ca, 5 wt.-% La and Ag bal.; Nanoval GmbH & Co. KG, Berlin, Germany) powder. As reference, pure FeMn and FeMn modified with 5 wt.-% pure Ag (ECKART TLS, Bitterfeld-Wolfen, Germany) are investigated. An amount of 5 wt.-% is selected, as the results of Huang et al. [39], Niendorf et al. [22], and Liu et al. [65] demonstrate enhanced degradation, biocompatibility, and acceptable mechanical properties for such an amount of Ag. The powders with a nominal particle size between 20 and 63 μm are produced via argon gas atomization. All powders have a suitable particle size distribution that is almost within the desired range and an almost spherical shape with only a few satellites [63,64]. Upright plates with a size of $2.5 \times 10 \times 12.5$ mm^3 are manufactured with an SLM 280^{HL} (SLM Solutions Group AG, Lübeck, Germany) in an argon atmosphere (layer thickness: 50 μm, scanning velocity: 750 mm/s, laser power: 280 W, hatch distance: 110 μm) [63,64]. The resulting FeMn matrix is characterized by an austenitic structure and segregations of Mn along with the melt-pool structure of LBM as well as the columnar dendritic solidification structure. The Ag and AgCaLa phases are homogeneously dispersed in the FeMn matrix and have a comparable morphology (approx. size: 30–370 μm). The Ag phases contain 88.3 wt.-% Ag, 11.7 wt.-% Mn, and 0.1 wt.-% O. Small slag particles are included in the AgCaLa phases. The AgCaLa phases contain 75.8 wt.-% Ag, 10.6 wt.-% Ca, 5.4 wt.-% La, 5.2 wt.-% Mn, and 3.9 wt.-% O. A detailed characterization of the LBM-processed materials and conventionally processed AgCaLa alloy has been carried out by Krüger et al. [41,63,64]. In the following, the samples are designated as FeMn, FeMnAg, and FeMnAgCaLa.

2.1. Electrochemical Investigations

A Potentiostat MLab200 (Bank Elektronik—Intelligent Controls GmbH, Pohlheim, Germany; Potential resolution: +/− 1 mV; Current resolution: 200 pA) with a three-electrode arrangement is used to determine the open circuit potential (OCP) and the linear sweep voltammetry (LSV). The reference electrode is an Ag/AgCl electrode, type SE11 in saturated KCl solution (Xylem Analytics Germany Sales GmbH & Co. KG, Weilheim, Germany), and the counter electrode is a platinum sheet. The samples are fixed in an inert sample holder with a circular contact surface of 24.4 mm^2. Ringer–Lactate solution (RL) (B. Braun SE, Melsungen, Germany) is used as the electrolyte. The samples are ground before the measurement with a grid size of P1200 to remove oxide layers. A stabilized state is achieved by waiting 5 min after arranging the set-up. During the OCP measurement, the data are recorded at 10 Hz for 10 min. The LSVs are collected with new samples between −1800 mV vs SHE and 1200 mV vs SHE, with a scan rate of 1 mV/s.

2.2. Scanning Kelvin Probe Force Microscopy (SKPFM)

SKPFM studies were performed to determine the contact potential differences between the matrix and the Ag phases. Samples were ground with SiC paper (grit P600, P1000, P2500, and P4000) and polished with 1 μm diamond paste (Schmitz-Metallographie GmbH, Herzogenrath, Germany). The polished samples were cleaned by sonication in ethanol for 10 min (Ultrasonic Cleaner, 45 kHz, 120 W, VWR International GmbH, Darmstadt, Germany). SKPFM measurements were performed using an MFP-3D-AFM setup (Oxford Instruments Asylum Research Inc., Santa Barbara, USA) with a platinum-coated tip (MikroMasch HQ:NSC15/Pt, 325 kH, 40 N m^{-1}, NanoAndMore GmbH, Wetzlar, Germany). The delta height was set to 40 nm. The scan rate was 0.1 Hz, and the amplitude was 1 V. A 3 V bias voltage was applied to the probe during the measurement. In this case, a higher potential difference indicates a more negative electrode potential [66]. Three parallel samples of each alloy were measured.

2.3. Immersion Tests

The immersion tests are performed for 56 days under varying conditions (Table 1). Five different electrolytes (Table 2) are used to investigate the influence of the specific components of the immersion solution and the pH value: Hanks' Solution (HS) with Ca++ and Mg++ (Lonza Group AG, Basel, Switzerland), RL, NaCl (NC, 0.9 %), pH3-, and pH5-solution. The solutions should imitate the conditions in the human body. For each combination, three samples are tested together in a laboratory glass bottle. One of three samples is ground iteratively up to P4000 before the immersion test to enable detailed investigations after immersion. The others are both used as build conditions. The samples have a size of 2.5 × 10 × 12.5 mm^3, and a ratio of 9.2 mL/cm^2 is used for immersion. Holes with a diameter of 1 mm in the samples enable the free-hanging positioning in the immersion solution with a polypropylene filament. The influence of solved gases is investigated by the application of additional bubbling air (0.2 l/h), increasing the exchange with the atmosphere compared to aeration. Furthermore, the influence of temperature is investigated at 20 °C and 37 °C.

Table 1. Combination of immersion conditions and solutions.

Solution \ Condition	Aeration 20 °C	Aeration 37 °C	Bubbled Air 20 °C	Bubbled Air 37 °C
Ringer–Lactate Solution (RL)	X	X	X	X
Hanks' Solution (HS)	X	X	X	X
NaCl Solution (NC)	X	X	X	X
pH5		X		
pH3		X		

The samples are inspected after 1, 3, 6, and 10 days and, subsequently, every 7 days. At each inspection, the samples are removed from the solution, and the solution is exchanged. After the removal, the samples are rinsed (slow movement in a bath) in Ringer rinsing solution (B. Braun SE, Melsungen, Germany) and acetone. The samples are dried with hot air for 30 s. This treatment removes loose degradation products which imitate the conditions in the human body, where a complete removal does not occur but rather some flow and exchange of the surrounding solution. The sample weight is measured with an XP205 scale (Mettler-Toledo LLC, Columbus, OH, USA) with an accuracy of 0.01 g, and images of the samples are taken. After the immersion test, the samples are investigated via Raman spectroscopy and field emission scanning electron microscopy (FE-SEM). Therefore, a Zeiss Ultra Plus (Carl Zeiss AG, Oberkochen, Germany) with a secondary electron (SE) detector is used. The energy-dispersive X-ray spectroscopy (EDS) detector Octane Pro (AMETEK,

Berwyn, PA, USA) of the SEM enables the determination of the chemical composition. Before investigating via SEM, the samples are cleaned for 30 s in an acetone ultrasonic bath.

Table 2. Composition of solutions utilized for static immersion tests.

Components		Concentration g/L				
		HS	RL	NC	pH5	pH3
Calcium Chloride Dihydrate	$CaCl_2 \bullet 2H_2O$	0.186	0.270	—	—	—
Dextrose	$C_6H_{12}O_6$	1.000	—	—	—	—
Magnesium Sulfate Heptahydrate	$MgSO_4 \bullet 7H_2O$	0.200	—	—	—	—
Potassium Chloride	KCl	0.400	0.400	—	—	—
Potassium Phosphate Monobasic Anhydrous	KH_2PO_4	6.000×10^{-2}	—	—	—	—
Sodium Bicarbonate	$NaHCO_3$	0.350	—	—	—	—
Sodium Chloride	NaCl	8.000	6.000	9.000	—	1.690
Sodium Phosphate Dibasic-7-Hydrate	$Na_2HPO_4 \bullet 7H_2O$	9.000×10^{-2}	—	—	—	—
Phenol Red	$C_{19}H_{14}O_5S$	2.000×10^{-2}	—	—	—	—
Sodium Lactate	$NaC_3H_5O_3$	—	3.120	—	—	—
Citric Acid	$C_6H_8O_7$	—	—	—	0.940	3.760
Caustic Soda	NaOH	—	—	—	0.400	0.400

The degradation behavior of the AgCaLa phases after release from the FeMn matrix is investigated with an immersion test over 6 months to enable the dissolution of the matrix and the release of Ag and AgCaLa particles, respectively. Half of the RL is changed every two weeks in order to guarantee the remaining released particles in the immersion vessels. To dissolve depositions after the immersion test, ultrasonic is applied for 30 s. The solution is removed from the immersion vessel, and the residues and samples are dried and investigated via SEM.

2.4. Raman Spectroscopy

Raman spectra were measured to identify corrosion products by using an InVia Renishaw Raman microscope (Renishaw plc, Wotton-under-Edge, UK) with a (charge-coupled device) CCD detector. A 532 nm yttrium-aluminum-garnet (YAG) laser, an 1800-L/mm grating and a 50× objective were used. The laser power at the sample was adjusted to 0.1 mW (power densities: 0.2 mW/μm^2). The energy resolution was 0.3 cm^{-1} (full width at half maximum, FWHM).

2.5. Cytocompatibility Tests and Microbiological Analyses

The experiments are carried out with the murine fibroblast cell line L-929 (CLS Cell Lines Service GmbH, Eppelheim, Germany) and the human osteosarcoma cell line 87070202 (HOS; European Collection of Authenticated Cell Cultures Merck GmbH, Darmstadt, Germany). Fibroblasts and osteosarcoma cells were chosen since implanted alloys may come in contact with these cell types. Furthermore, L929 cells are recommended according to DIN ISO 10933-5 for viability testing. L-929 cells are grown and passaged in Roswell Park Memorial Institute (RPMI)-1640 medium (Biochrom GmbH, Berlin, Germany) supplemented with 10% fetal calf serum (FCS) superior (Biochrom GmbH, Berlin, Germany) and 1% penicillin/streptomycin (Pen/Strep) (Biochrom GmbH, Berlin, Germany). Eagle's Minimum Essential Medium (EMEM)/Hanks' (Carl Roth GmbH + Co. KG, Karlsruhe, Germany) maintained 10% FCS, 1% Pen/Strep, and 1% non-essential amino acids (Biochrom GmbH,

Berlin, Germany), and 2 mM L-glutamine (Biochrom GmbH, Berlin, Germany) was used for HOS cells. Both cell lines are grown and passaged in cell culture flasks or multi-well plates (Greiner Bio-One GmbH, Frickenhaus Germany). A 0.05% trypsin/0.02% ethylene-diamine-tetraacetic acid solution (Biochrom GmbH, Berlin, Germany) is used for passaging. The cells are plated with a density of 10,000 cells per well in 96-well microtiter plates. For experimental exposure, each alloy is immersed in 10 mL of the cell culture medium or Mueller–Hinton broth (Carl Roth GmbH + Co. KG, Karlsruhe, Germany) without additives. After a 3-day incubation at 37 °C in a humidified atmosphere of 5% CO_2, the cells undergo 24 h of incubation with the degradation medium in a humidified atmosphere at 37 °C and 5% CO_2.

2.6. Cell Viability Test

Cell viability is determined by a 3-(4,5-dimethylthiazol-2-yl)-5-(3-carboxymethoxyphenyl)-2-(4-sulfophenyl)-2H-tetrazolium inner salt (MTS) assay (CellTiter 96® Aqueous One Solution Cell Proliferation Assay, Promega Corp., Madison, WI, USA) with confluent cells after incubating with degradation media for 24 h, as previously described by Krüger et al. [41]. Each concentration is measured eight times, while each experiment is performed six times.

2.7. Test for the Antibacterial Activity of the Degradation Supernatant

Standard strains of *Escherichia coli* (*E. coli*; ATCC® 25922, Leibniz Institute DSMZ-German Collection of Microorganisms and Cell Cultures GmbH, Braunschweig, Germany) are used since these bacteria are common etiologic agents of clinical infections involved in bone tissue infections related to implants. The frozen stock is incubated at 37 °C for 24 h on sheep blood agar plates (Thermo Fisher Scientific Inc., Waltham, MA, USA) before starting the experiments.

The in vitro exposure of *E. coli* to degradation media (as described above) is performed using 96-well microtiter plates with 10^4 colony-forming units (cfu) per well. Each 96-well microtiter plate contains a set of six aliquots of every sample containing bacteria. After 24 h of incubation at 36 °C ± 1 °C, the optical density (OD) of each well of a microtiter plate is measured with a photometer at 450 nm. Each experiment is performed four times.

3. Results

The electrochemical analysis of the corrosion current densities as a function of the applied electrode potential (LSV curves) in RL is illustrated in Figure 1, along with the transient of the free corrosion potential open circuit potential (OCP) over the immersion time. The measured data indicate a change in the corrosion kinetics as a function of the alloy composition. The LSV data of FeMn and FeMnAg did not show any significant difference in the anodic and cathodic regions (Figure 1). In agreement with this result, the corresponding OCP data of FeMn and FeMnAg showed a similar trend.

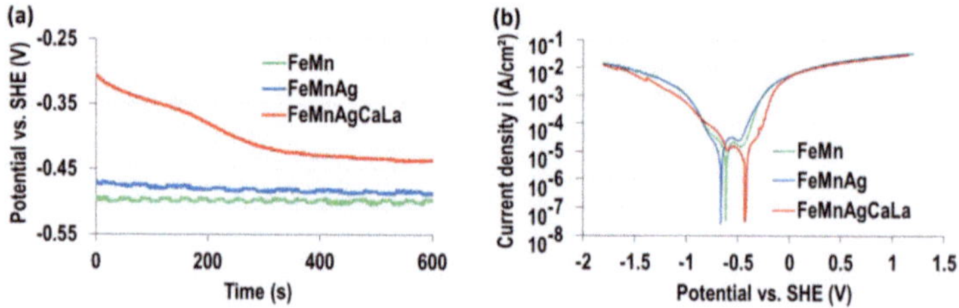

Figure 1. Electrochemical properties determined in RL: (**a**) transient of free corrosion potential; (**b**) linear sweep voltammetry.

The initial OCP values during the first 10 min of immersion (Figure 1a) show a more anodic corrosion potential for the FeMnAgCaLa alloy in comparison to the FeMn and FeMnAg alloy substrates. This result corresponded well with the corresponding LSV data showing that that the anodic dissolution rate is reduced for the FeMnAgCaLa alloy in comparison to the two other alloys.

At high anodic overpotentials, the current densities of all three alloys become similar, which can be explained by the dominating anodic dissolution of the FeMn phase in all cases.

Figure 2 shows the AFM topographic and volta potential maps of polished FeMnAg (Figure 2a–c) and FeMnAgCaLa (Figure 2d–f). The SKPFM data clearly show a potential difference between the two main phases of FeMnAg (Figure 2a) and FeMnAgCaLa (Figure 2d).

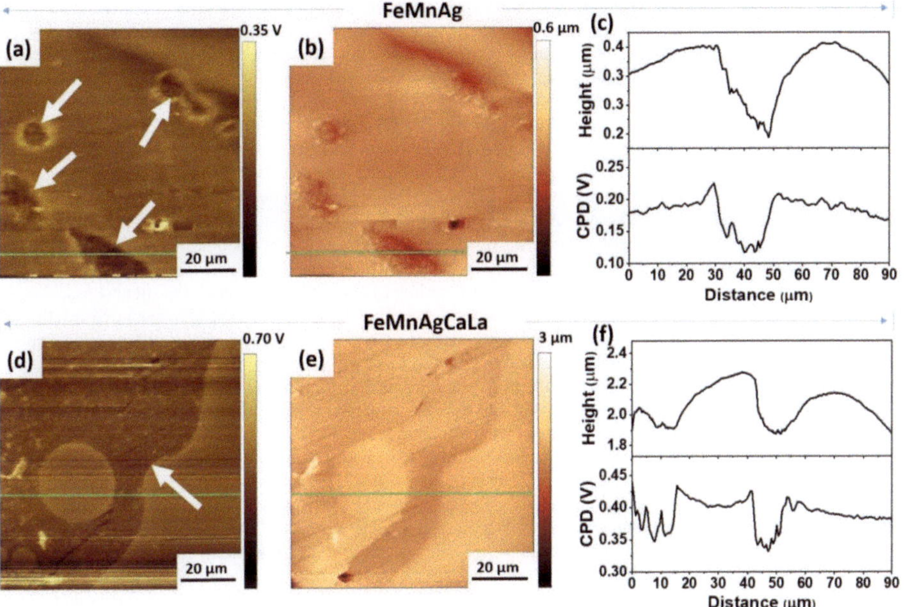

Figure 2. SKPFM results of FeMnAg and FeMnAgCaLa obtained from a 90 μm × 90 μm area: (**a,d**) topography, (**b,e**) corresponding contact potential differences, (**c,f**) corresponding cross-section (green line) profiles.

The Ag and AgCaLa phases (white arrows) are darker, indicating a lower CPD value of Ag phases, which confirms the more noble character compared to the FeMn phase. The cross-section profiles (green line in Figure 2) determined by SKPFM indicate a difference of about 70 mV between the CPD values of the Ag and AgCaLa phases and the FeMn phase. In the same interval, the height profile determined by AFM shows that the Ag and AgCaLa phases form grooves on the surface after the polishing step due to their lower hardness. In addition, FeMnAgCaLa obtained a higher root mean square (rms) roughness value of 129.3 ± 49.2 nm compared to that of FeMnAg (48.8 ± 17.5 nm).

During the immersion tests, the course of mass removal is similar for all three investigated materials under comparable conditions (Figure 3a). Thus, the expected increased degradation due to the addition of Ag and AgCaLa does not occur. The same results are obtained for all combinations of immersion conditions. The observation of the sample surface confirms the absence of a significant effect of the Ag and AgCaLa phases, as no crucial impact of the Ag phases on the FeMn is observed (Figure 3b,d). Just a small gap with a dimension of 1–3 µm around the Ag and AgCaLa phases has emerged (Figure 3d, white arrows). Such gaps have not been present before degradation [64]. The Ag and AgCaLa phases remain, while the matrix is dissolved, as expected. The structure of the melt pools (Figure 3b) and the structure of columnar dendritic solidification (Figure 3c) are observable. Thus, the structure of segregations of Mn along the microstructure observed by Krüger et al. [63] becomes visible during the immersion test, and the influence of segregations on the degradation is conclusive.

The degradation proceeds continuously. For long-time immersion tests over 6 months, continuous degradation is observed. A few more depositions are observed after 1344 h (Figure 3d) compared to 700 h (Figure 3c). These depositions do not significantly affect the degradation since mass removal proceeds continuously. Thus, a continuous transfer of function from the implant back to the human organism might be possible.

The pH value influences the degradation process significantly, as the removal is approximately 100 times higher than that for neutral pH values (Figure 3e). The removal at pH3 is higher than that at pH5. An increased degradation rate of FeMnAgCaLa compared to FeMn is observed at pH3 and pH5. An increased degradation due to the addition of pure Ag is observed at pH3. The course of mass removal is conclusive with the samples after immersion (Figure 3f–h), since the significantly increased removal of the matrix around the AgCaLa phase is observed. Thus, the expected increased degradation by modification with Ag and AgCaLa occurs for low pH values. As expected, the Ag and AgCaLa phases remain while the matrix degrades. This, together with the increased degradation of the matrix, indicates a higher corrosion resistance of Ag and AgCaLa and the effectiveness of a higher electrochemical potential for low pH values.

Aside from the pH value, the composition of the immersion solution, temperature, and intensity of exchange with the atmosphere affect the degradation rate. The curves in Figure 3i–k are the average material removal (FeMnAg) during immersion under varying conditions. As expected, higher temperatures and additional bubbling air increase the degradation rate (Figure 3j,k). Bubbling air intensifies the exchange with the air, enhancing the uptake of necessary O, and higher temperatures promote the chemical reactions. As for immersion in HS (Figure 3i), the weight is almost constant compared to immersion in RL and NC, and the components of the immersion solution have a significant influence. The constant weight during immersion in HS is conclusive since only slight depositions are observed on an almost unchanged sample surface (Figure 3n), whereas the sample surfaces after immersion in RL and NC are characterized by the removal of matrix material and depositions (Figure 3l,m). These depositions are more pronounced for immersion in RL, which might be correlated with the slightly lower mass removal compared to immersion in NC.

The degradation proceeds continuously over time but heterogeneously over the sample surface. The FeMnAgCaLa sample shown in Figure 4 is covered by brown depositions in the upper corners after immersion. The rest, except for the AgCaLa phases, is covered by a black deposition (Figure 4a). These depositions are removable by ultrasonic cleaning (Figure 4b). The surface is characterized by the significant corrosion removal of the FeMn matrix in the lower area (Figure 4c) and slight corrosion in the upper corners (Figure 4d). The brown depositions might influence further degradation.

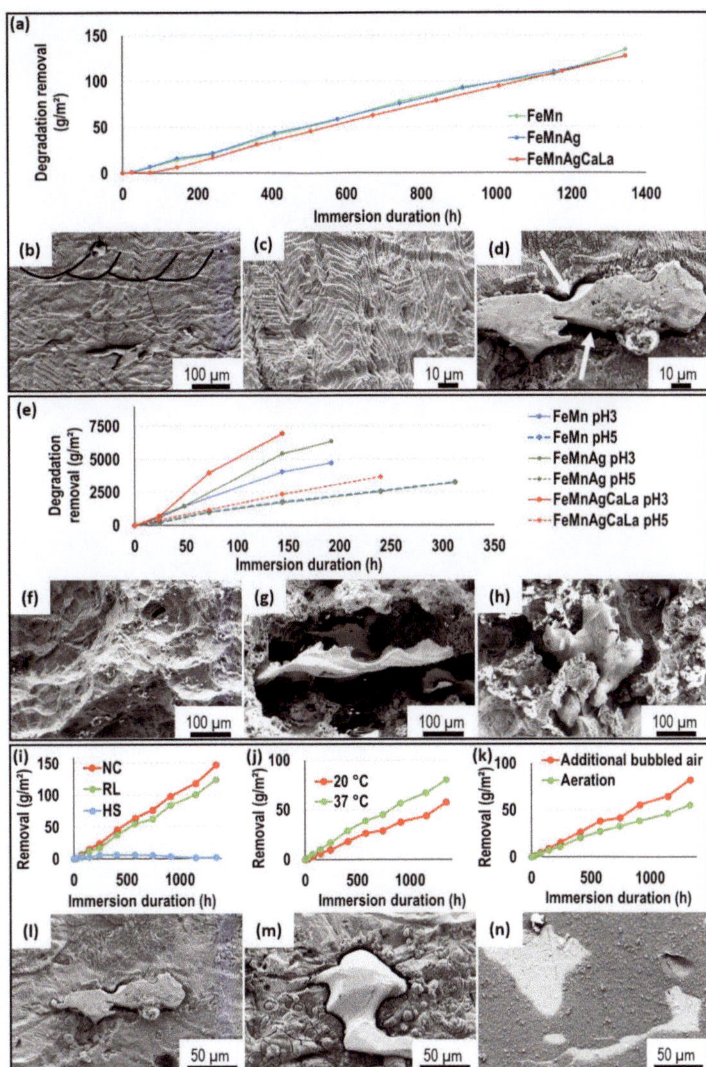

Figure 3. (**a**) Progress of degradation depending on material (N = 3, 37 °C, RL, aeration), SEM-SE: (**b**) FeMnAg after 700 h, (**c**) FeMn after 700 h, (**d**) FeMnAg after 1344 h; (**e**) progress of degradation depending on pH value (N = 3, 37 °C, aeration), SEM-SE: (**f**) FeMn after 192 h at pH3, (**g**) FeMnAg after 192 h at pH3; (**h**) FeMnAgCaLa after 144 h at pH3; progress of degradation of FeMnAg depending on (**i**) immersion solution (N = 12), (**j**) temperature (N = 18), (**k**) exchange with atmosphere (N = 18), SEM-SE after 1344 h: (**l**) FeMnAg (37 °C, NC, aeration), (**m**) FeMnAg (37 °C, RL, aeration), (**n**) FeMnAg (37 °C, HS, aeration).

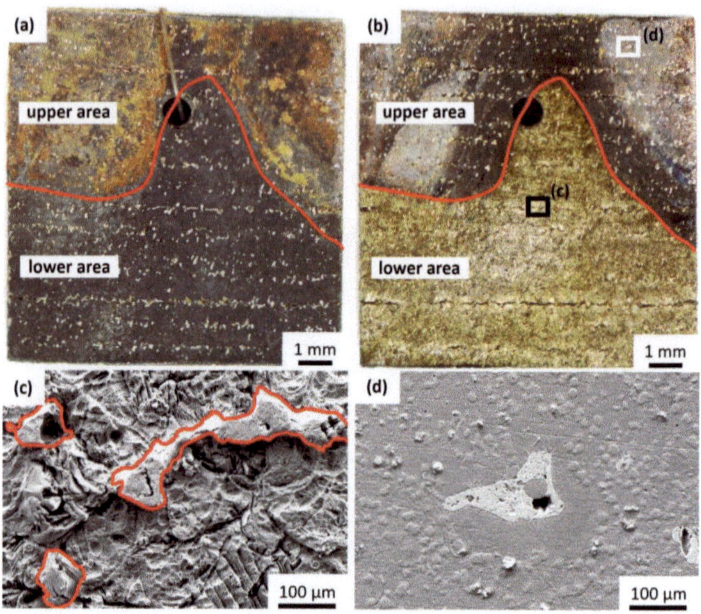

Figure 4. Local differences in degradation progress, FeMnAgCaLa after 1344 h (37 °C, RL, aeration): (**a**) image before ultrasonic cleaning; (**b**) image after ultrasonic cleaning; (**c**) SEM-SE: Magnified section of black box in (**b**); (**d**) SEM-SE: Magnified section of white box in (**b**).

The AgCaLa phases (Figure 4c) are characterized by a more fragmented structure with pores compared to the Ag phases (Figure 3d). This structure is in accordance with the structure of AgCaLa phases before degradation, since some porosity in connection to the AgCaLa phases as well as inclusions of slag are observed. The dissolution of these inclusions is expectable and results in finely structured porosity [64].

The Raman spectra of samples after degradation in RL, pH5, and pH3 electrolytes are shown in Figure 5. The Raman spectra were assigned to specific microscopically identified characteristic regions.

The Raman spectra of the FeMnAg and FeMnAgCaLa samples for the FeMn phase (blue circles in Figure 5a–f) and the Ag phase (red circles in Figure 5a–f) are shown in Figure 5g,h. For the FeMn phase, peaks were observed at 213 cm^{-1}, 279 cm^{-1}, 389 cm^{-1}, 610 cm^{-1}, and 1280 cm^{-1}, which are characteristic of α-Fe$_2$O$_3$ [67]. The most relevant peaks in the spectra of the FeMn phases of both samples are similar for all pH conditions. In the areas of Ag and AgCaLa phases, significant broad absorption bands at around 600 cm^{-1} were observed. This peak might be assigned to Fe$_3$O$_4$ or Ag-O stretching and bending modes, respectively [42,68]. The band at 1300 cm^{-1} is assigned to the carboxylic acid group and can be explained by the lactic acid in the RL solution. Finally, the band at around 1550 cm^{-1} is assigned to AgCl [69]. The results of Raman spectroscopy showed that similar surface layers were formed on FeMnAg and FeMnAgCaLa alloys immersed at different pH values. However, the spectra support the assumption that the Ag and Ag-rich phases act as local cathodes during the corrosion process, as already predicted based on the SKPFM mappings.

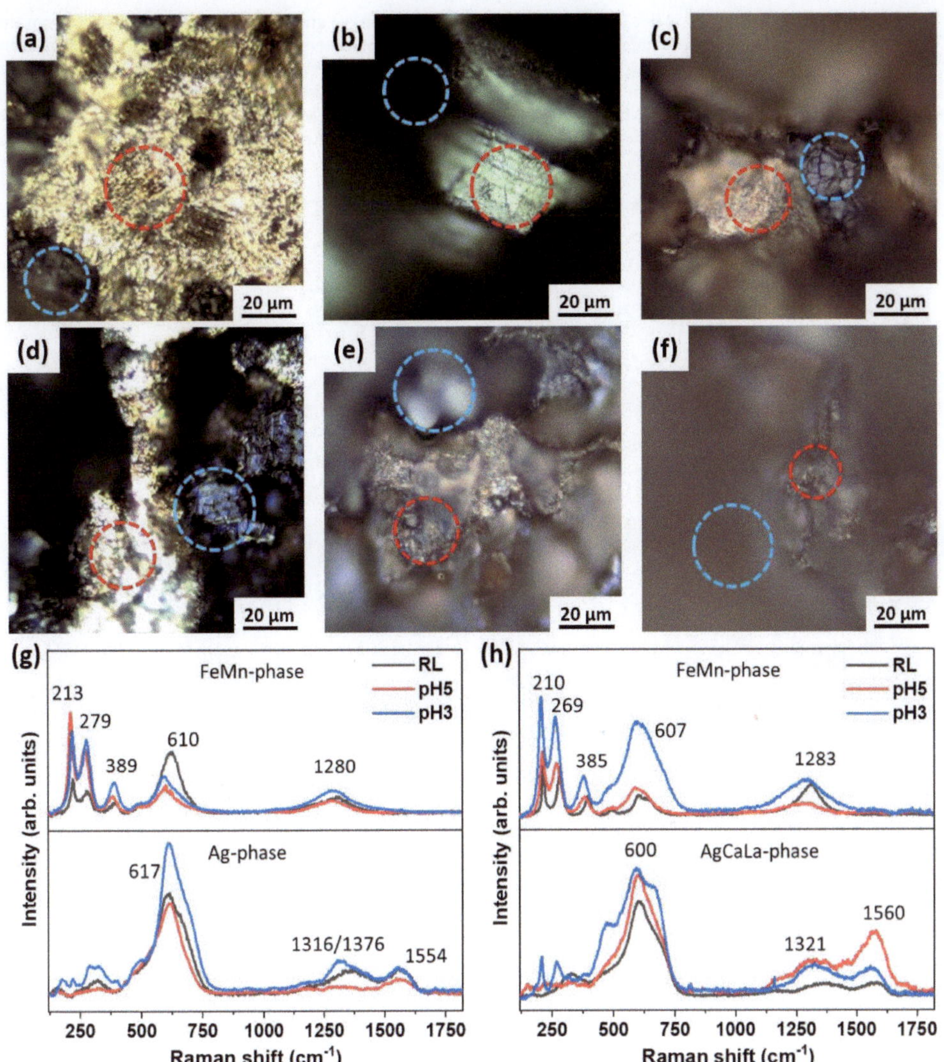

Figure 5. Samples after degradation (37 °C, aeration): Light microscopy image of FeMnAg: (**a**) 1244 h, RL, (**b**) 312 h, pH5, (**c**) 192 h, pH3; Light microscopy image of FeMnAgCaLa: (**d**) 1244 h, RL, (**e**) 240 h, pH5, (**f**) 144 h, pH3; (**g,h**) Raman spectra of the Ag and AgCaLa phase (red circles) and the FeMn phase (blue circles) at different pH values.

After immersion over six months, the Ag and AgCaLa particles released from the matrix are observed in the residues in the immersion vessel (Figure 6a,b). Significantly more particles are identified via SEM-EDS in the residues of FeMnAg (Figure 6a) than in the residues of FeMnAgCaLa (Figure 6b). Additionally, much more remaining Ag is observed on the FeMnAg sample (Figure 6c) than on the FeMnAgCaLa sample (Figure 6d). As both materials contain a similar amount of Ag and AgCaLa before immersion, the AgCaLa dissolves.

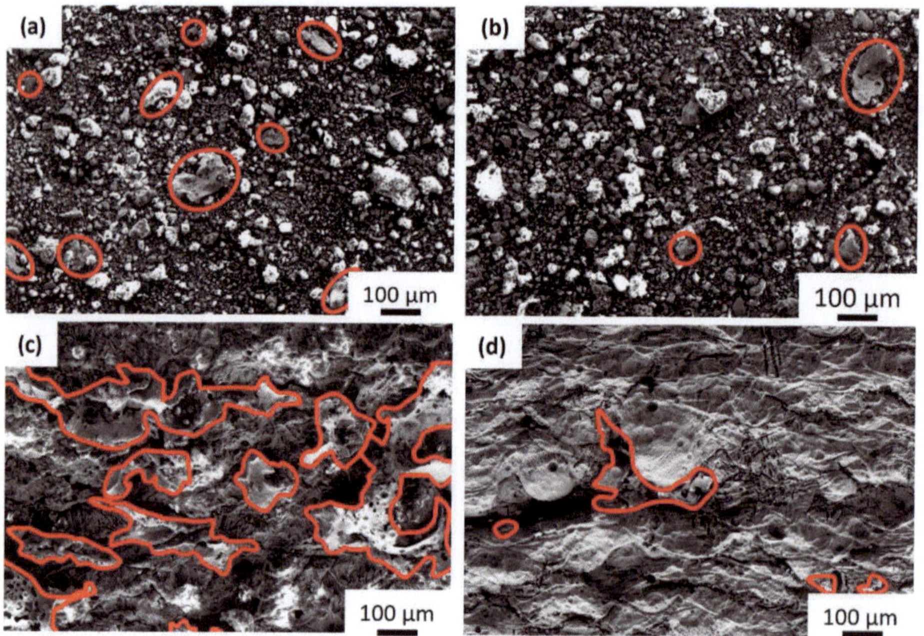

Figure 6. Degradation of Ag phases after 6-month immersion (37 °C, RL, aeration), SEM-SE: Particles remaining at the bottom of the immersion vessel; Ag identified via EDS and red-marked: (**a**) FeMnAg, (**b**) FeMnAgCaLa. Samples after immersion, particles marked with red border: (**c**) FeMnAg, (**d**) FeMnAgCaLa.

After 24 h in degradation media, the cell viability is affected in each case (Figure 7). Reduction below the threshold of 70% (DIN ISO 10993-5) is found in the HOS cells for viability in the case of FeMnAgCaLa. A similar reduction is demonstrated for FeMn samples. Significant differences can be found between FeMnAg and FeMnAgCaLa and FeMn, respectively.

The degradation media of FeMnAgCaLa led to a significant reduction in bacterial growth compared to the reference, while FeMn and FeMnAg induced only a slight reduction in bacterial growth (Figure 8).

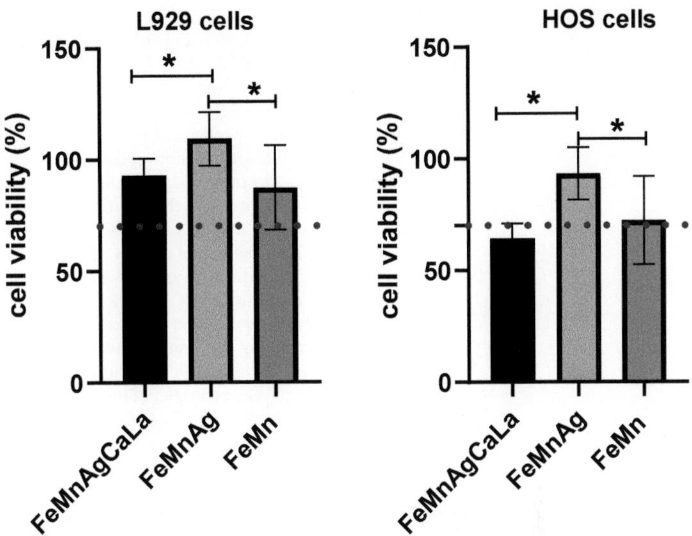

Figure 7. Effects of incubation with degradation media on the viability of L-929 and HOS cells after 24 h; the red dotted line demonstrates the lower limit at 70% to the reduced cell viability according to DIN ISO 10933-5; mean $\pm$ SD, n = 6 with 6 technical replicates, Kruskall–Wallis test with Dunn's multiple comparisons test; $p < 0.005 = *$ (GraphPad Prism 9.1.0.; GraphPad Software, LLC, San Diego, Cl, USA).

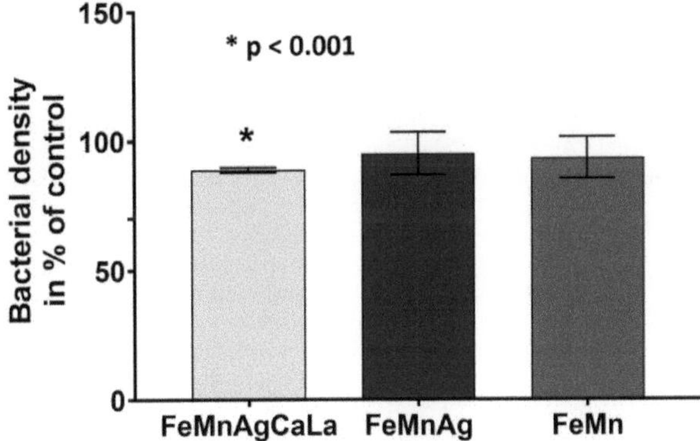

Figure 8. Effect of degradation supernatant on the growth of the reference strain *Escherichia coli* (ATCC® 25922). Results represent effects caused by the previous degradation in 3 mL Mueller–Hinton broth. Mean value $+/-$ standard deviation from four replicates. The optical density of the untreated control was set as 100%. Repeated measures analysis of variance (ANOVA) summary with Holm–Šídák's multiple comparisons test in comparison to control; $p < 0.001 = *$ (GraphPad Prism 9.1.0.; GraphPad Software, LLC, San Diego, CL, USA).

4. Discussion

The results prove the potential of FeMn modified with AgCaLa as resorbable material, since the degradation of AgCaLa after the dissolution of the FeMn matrix and biocompatibility are observed. Nevertheless, the cell viability is reduced below the threshold of 70% (DIN ISO 10993-5) in HOS cells treated with AgCaLa. However, this result is negligible, as FeMn and FeMnAg affect the cell viability as well, and FeMn is biocompatible (cf. introduction) [13,19,27–31]. As intended, an antibacterial effect is caused by FeMnAgCaLa, probably due to released Ag ions forming the degradable AgCaLa. Additionally, the fundamental proof of concept is successful, since the addition of Ag and AgCaLa enhances the degradation rate at low pH values, and small gaps surrounding Ag and AgCaLa are observed at neutral pH values. Some challenges remain, since the degradation rate is not increased at neutral pH values. The ineffectiveness of Ag and AgCaLa is confirmed by the similarity of LSV and the identical higher OCP of the modified FeMn, indicating similar and/or lower degradation rates. An in-depth analysis of the degradation behavior is necessary to achieve an understanding of the degradation mechanisms and enable the identification of the effect(s) leading to a lower anodic dissolution than theoretically possible. Based on this, adaptions of the material can be implemented.

To explain the ineffectiveness of Ag, the absence of a potential difference between Ag phases and the matrix would be a possible explanation, since the pure Ag takes up Mn from the matrix during LBM, and AgCaLa is affected by the inclusion of slag [63,64]. The SKPFM proves the higher potential of the Ag and AgCaLa phases compared to the matrix. In addition, the higher resistance of Ag and AgCaLa against corrosion is confirmed by the remaining Ag and AgCaLa, while the FeMn matrix dissolves. Thus, the absence of a potential difference is not reasonable for the ineffectiveness of the Ag and AgCaLa.

A further explanation might be the formation of corrosion-inhibiting layers due to the deposition of corrosion products. This hypothesis is supported by the depositions observed on the samples. The detection of O and, in particular, P and Cl via EDS on the Ag and AgCaLa phases is a hint of the presence of heavily dissolvable Ca-P and Ag-Cl compounds, suppressing further degradation [13,14,19,30,36,70]. Tonna et al. [14] reported a reduced degradation rate for Hank's solution containing Ca in contrast to Hank's solution containing no Ca. Thus, the degradation removal is reduced in case of possible Ca-P formation. This is following the results presented in this study. However, for the presented results, the formation of Ca-P is limited by the absence of P in NC and RL and not the absence of Ca. The Raman spectra confirm the presence of Ag-Cl on Ag and AgCaLa. The hypothesis of inhibiting depositions is supported by the formation of gaps around the Ag and AgCaLa phases. No depositions exist at the beginning of degradation, and the potential difference can effectively increase the dissolution of the matrix (FeMn), resulting in gaps [71]. During the development of the gaps, the inhibiting layer develops, and further anodic dissolution does not occur. Gaps around Ag phases were observed by Dargusch et al. [35] in combination with an increased degradation rate. For the presented results, the influence of the gaps is too small, and the weight of the remaining Ag phases counteracts an effective increase in degradation. Furthermore, the effectiveness of Ag and AgCaLa at low pH values is in accordance with the blocking effect, since the solubility of degradation products increases at low pH values, and the absence of inhibiting layers at low pH values is conclusive [44,72,73].

However, the ineffectiveness of Ag and AgCaLa for neutral pH values is unexpected, since an increasing effect due to anodic dissolution should occur and is reported in the literature [22,31,35,39,42,74]. An increased degradation removal after 7 days of immersion in 0.9 % NaCl solution of additively processed FeMn with 5 wt.-% Ag almost identical to the material investigated in this study was reported by Niendorf et al. [22]. The Ag phases should suppress the formation of degradation-inhibiting layers [44]. However, for the presented results, this effect does not occur, since Ca, P, and Ag-Cl are identified on FeMn and Ag, indicating the deposition of inhibiting layers. Wang et al. [44] assumed that the cathodic reaction releasing hydroxyl ions might support the deposition on the

Ag phases. This hypothesis follows the reported ineffectiveness of phases with high potential [29,75–77]. Loffredo et al. [75] observed a slight reduction in degradation removal after immersion in Hank's modified solution for 14 days due to modification with Ag. They suggest that the earlier formation of inhibiting layers is the reason.

Thus, the results and explanations for the degradation behavior reported in the literature differ, and the obtained results are in accordance with some studies [78]. The differences between the studies are crucial to explaining the discrepancy of degradation and identifying the decisive influencing properties in order to adapt them. As the presented results demonstrate, the immersion conditions are more decisive than the chemical composition of the materials. The results of Pierson et al. [33] underline this relationship, since material implanted into the arterial wall degrades, whereas no corrosion occurs in the bloodstream for the same material. The local differing degradation, which is also reported in the literature, emphasizes the sensitivity to immersion conditions, since the differences are caused by the local variation of conditions over the sample surface [36,79]. Accordingly, the influence of the immersion conditions is considered below.

A significant difference between the degradation in NC, RL, and HS is present. HS is the only solution containing P, and, thus, Ca-P compounds may form layers that significantly affect further degradation [13,19,28,30,36,45,71]. Therefore, the degradation in HS might be suppressed by the deposition of compounds originating from the solution. The reproduction of the significant mass removal in NC and RL, which might be suitable for some applications, is not expectable in vivo due to the presence of P, but the presence of P is not suitable as the only explanation for the ineffectiveness, since, in the literature, degradation is reported for immersion in P- and Ca-containing solutions and in in vivo experiments with the presence of Ca and P [12,13,19,35,38,39,65,79,80]. Accordingly, further aspects have to be considered. Dryndra et al. [28], Kraus et al. [29], and Peuster et al. [18] reported no significant degradation in the absence of significant inflammation, whereas Mandal et al. [38], Feng et al. [32], and Dargusch et al. [35] reported degradation in combination with an inflammatory response. These results are conclusive, since the pH value is reduced when an inflammatory response occurs, and a reduced pH value significantly affects the degradation rate, as the results presented in this study demonstrate [44,72,81]. The inflammatory response might explain the different results of in vivo studies. Otherwise, Lin et al. [16] report degradation together with slight inflammation. Thus, no single decisive parameter is identifiable for the ineffectiveness of Ag and AgCaLa. The degradation behavior results from a complex interaction of all characteristics of immersion conditions such as the pH value, the presence of P, and the material properties.

However, the results and explanations for the degradation behavior differ in the literature. Many influences are crucial, and, thus, the secure identification of reasons for the observed results is challenging. The drawn conclusions are based on suggestions about the most important influences. Even for this study, the conclusions are based on the experimental results linked with the information from the literature. Some contradictions are present, and this must be considered for the judgment of the conclusions drawn from the discussion.

In summary, the immersion conditions are crucial for the degradation behavior, and expected anodic dissolution might occur, as reported in the literature, under specific immersion conditions. However, for the successful application of the material, a reliable dissolution of the implant under all conditions would be favorable. The degradation rate must be suited regardless of an eventually occurring inflammatory response. It is challenging to fulfill these requirements since the results presented demonstrate the strong impact of the immersion conditions. Thus, the development of a material adapted for many possible applications is not expectable. Since the suitability of the material and the increased degradation by anodic dissolution might occur at some implantation sites, focusing on a more detailed simulation of conditions in the human body is appropriate. Although the degradation rate is not increased under neutral pH values, further investigations are needed, since the effectiveness of the concept of anodic dissolution by modification with

Ag, in general, is proven. The degradation rate is increased for low pH values, and gaps develop around Ag for neutral pH values. Furthermore, an adaption of the material is appropriate to enable the application for more use cases. A possibility is to introduce more, smaller AgCaLa phases to increase the impact of the gaps around the Ag phases. In addition, the formation of Mn segregations is promising, since a significant influence of segregations is observed. However, inhibiting layers are developed. Thus, the development of an additional mechanism to prevent the deposition of such layers is required to enable the effectiveness of potential differences [71].

5. Conclusions

The results presented demonstrate the impact of the modification of FeMn with Ag and AgCaLa, a degradable Ag-alloy, on the biocompatibility, antibacterial properties, and degradation behavior. Furthermore, an analysis of the degradation behavior of the FeMn matrix and the Ag phases identifies crucial factors of the immersion conditions and enables further investigations and adaptions of the material. The key findings can be summarized as follows:

- For low pH values, the addition of Ag increases the degradation rate.
- For neutral pH values, small gaps surrounding the Ag and AgCaLa phases develop.
- For immersion conditions adapted to the human body, no significantly increased degradation rate occurs.
- The Ag phases effectively cause anodic dissolution, but depositions of inhibiting layers suppress an increased degradation rate for the selected immersion conditions.
- The degradation behavior strongly depends on the pH value and the composition of the immersion solution—particularly, the presence of P.
- Mn segregations influence the local degradation behavior.
- Particles of the degradable AgCaLa alloy dissolve after the matrix material.
- Biocompatibility can be assumed for all materials.
- An antibacterial effect is observed for FeMn modified with degradable AgCaLa.

In summary, the results demonstrate the possibility of increasing the degradation rate by the addition of phases with high electrochemical potential, causing anodic dissolution. The results also reveal remaining challenges, as, for the selected immersion conditions with a neutral pH value, the degradation rate is not increased. Since high sensitivity to the immersion conditions is observed, the investigated material might be effective on some implantation sites. Accordingly, further research should focus on a more detailed simulation regarding the immersion conditions. Furthermore, an adaption of the material with smaller Ag phases or applying the influence of Mn segregations should be aimed for. To enable the effectiveness of the actual material, a mechanism to suppress the formation of inhibiting layers needs to be developed.

Author Contributions: Investigation, J.T.K., J.H., V.F., R.H.M.-V. and H.O.; writing—original draft preparation, J.T.K., J.H., H.O. and J.M.; writing—review and editing, J.T.K., K.-P.H., J.H., V.F., R.H.M.-V., J.M., G.G. and M.S.; supervision, K.-P.H., J.M., G.G. and M.S.; project administration, K.-P.H., J.M., G.G. and M.S.; funding acquisition, J.M., G.G. and M.S. All authors have read and agreed to the published version of the manuscript.

Funding: This research was funded by the German Research Foundation (DFG) cooperation project 414365989 (SCHA1484/44-1, KI 361/8-1, and GR 1709/27-1).

Institutional Review Board Statement: Not applicable.

Informed Consent Statement: Not applicable.

Data Availability Statement: Not applicable.

Acknowledgments: The authors gratefully thank the German Research Foundation (DFG) for the financial support. This work results from the cooperation project 414365989 (SCHA1484/44-1, KI 361/8-1, and GR 1709/27-1). Furthermore, the authors would like to thank Malte Dreyer, Sabrina Beumer, Eva Mühlhoff, and Lea Kaspersmeier for promoting the production and examination of the tested material. In addition, the authors thank Maxwell Hein and Steven Clifford Woodcock for proofreading, as well as Carina Dillgart, Caroline Groß, and Viktoria Nepke for the technical help with the cell culture experiments and the microbiological examinations.

Conflicts of Interest: The authors declare that they have no conflict of interest.

References

1. Chen, Q.; Thouas, G.A. Metallic implant biomaterials. *Mater. Sci. Eng. R Rep.* **2015**, *87*, 1–57. [CrossRef]
2. Zivic, F.; Affatato, S.; Trajanovic, M.; Schnabelrauch, M.; Grujovic, N.; Choy, K.L. *Biomaterials in Clinical Practice*; Springer: Cham, Switzerland, 2018.
3. Carluccio, D.; Demir, A.G.; Bermingham, M.J.; Dargusch, M.S. Challenges and Opportunities in the Selective Laser Melting of Biodegradable Metals for Load-Bearing Bone Scaffold Applications. *Metall. Mater. Trans. A* **2020**, *51*, 3311–3334. [CrossRef]
4. Ratner, B.D.; Hoffman, A.S.; Schoen, F.J.; Lemons, J.E. *Biomaterials Science*, 3rd ed.; Academic Press: Cambridge, MA, USA, 2013.
5. Veerachamy, S.; Yarlagadda, T.; Manivasagam, G.; Yarlagadda, P.K. Bacterial adherence and biofilm formation on medical implants: A review. *Proc. Inst. Mech. Eng. Part H J. Eng. Med.* **2014**, *228*, 1083–1099. [CrossRef] [PubMed]
6. Vignesh, M.; Kumar, G.R.; Sathishkumar, M.; Manikandan, M.; Rajyalakshmi, G.; Ramanujam, R.; Arivazhagan, N. Development of Biomedical Implants through Additive Manufacturing: A Review. *J. Mater. Eng. Perform.* **2021**, *30*, 4735–4744. [CrossRef]
7. Prakasam, M.; Locs, J.; Salma-Ancane, K.; Loca, D.; Largeteau, A.; Berzina-Cimdina, L. Biodegradable Materials and Metallic Implants—A Review. *J. Funct. Biomater.* **2017**, *8*, 44. [CrossRef] [PubMed]
8. Alonzo, M.; Primo, F.A.; Kumar, S.A.; Mudloff, J.A.; Dominguez, E.; Fregoso, G.; Ortiz, N.; Weiss, W.M.; Joddar, B. Bone tissue engineering techniques, advances and scaffolds for treatment of bone defects. *Curr. Opin. Biomed. Eng.* **2021**, *17*, 100248. [CrossRef]
9. Zhang, E.; Zhao, X.; Hu, J.; Wang, R.; Fu, S.; Qin, G. Antibacterial metals and alloys for potential biomedical implants. *Bioact. Mater.* **2021**, *6*, 2569–2612. [CrossRef]
10. Hermawan, H. Updates on the research and development of absorbable metals for biomedical applications. *Prog. Biomater.* **2018**, *7*, 93–110. [CrossRef]
11. Redlich, C.; Schauer, A.; Scheibler, J.; Poehle, G.; Barthel, P.; Maennel, A.; Adams, V.; Weissgaerber, T.; Linke, A.; Quadbeck, P. In Vitro Degradation Behavior and Biocompatibility of Bioresorbable Molybdenum. *Metals* **2021**, *11*, 761. [CrossRef]
12. Dargusch, M.S.; Dehghan-Manshadi, A.; Shahbazi, M.; Venezuela, J.; Tran, X.; Song, J.; Liu, N.; Xu, C.; Ye, Q.; Wen, C. Exploring the Role of Manganese on the Microstructure, Mechanical Properties, Biodegradability, and Biocompatibility of Porous Iron-Based Scaffolds. *ACS Biomater. Sci. Eng.* **2019**, *5*, 1686–1702. [CrossRef]
13. Mandal, S.; Viraj; Nandi, S.K.; Roy, M. Effects of multiscale porosity and pore interconnectivity on in vitro and in vivo degradation and biocompatibility of Fe–Mn–Cu scaffolds. *J. Mater. Chem. B* **2021**, *9*, 4340–4354. [CrossRef] [PubMed]
14. Tonna, C.; Wang, C.; Mei, D.; Lamaka, S.V.; Zheludkevich, M.L.; Buhagiar, J. Biodegradation behaviour of Fe-based alloys in Hanks' Balanced Salt Solutions: Part I. material characterisation and corrosion testing. *Bioact. Mater.* **2022**, *7*, 426–440. [CrossRef] [PubMed]
15. Abbaspour, N.; Hurrell, R.; Kelishadi, R. Review on iron and its importance for human health. *J. Res. Med. Sci.* **2014**, *19*, 164–174. Available online: https://www.ncbi.nlm.nih.gov/pmc/articles/PMC3999603/ (accessed on 17 January 2022). [PubMed]
16. Lin, W.; Qin, L.; Qi, H.; Zhang, D.; Zhang, G.; Gao, R.; Qiu, H.; Xia, Y.; Cao, P.; Wang, X.; et al. Long-term in vivo corrosion behavior, biocompatibility and bioresorption mechanism of a bioresorbable nitrided iron scaffold. *Acta Biomater.* **2017**, *54*, 454–468. [CrossRef]
17. Trumbo, P.; Schlicker, S.; Yates, A.A.; Poos, M. Dietary reference intakes for energy, carbohydrate, fiber, fat, fatty acids, cholesterol, protein and amino acids. *J. Am. Diet. Assoc.* **2002**, *102*, 1621–1630. [CrossRef]
18. Peuster, M.; Hesse, C.; Schloo, T.; Fink, C.; Beerbaum, P.; von Schnakenburg, C. Long-term biocompatibility of a corrodible peripheral iron stent in the porcine descending aorta. *Biomaterials* **2006**, *27*, 4955–4962. [CrossRef]
19. Hermawan, H.; Purnama, A.; Dube, D.; Couet, J.; Mantovani, D. Fe–Mn alloys for metallic biodegradable stents: Degradation and cell viability studies. *Acta Biomater.* **2010**, *6*, 1852–1860. [CrossRef]
20. Dehestani, M.; Trumble, K.; Wang, H.; Wang, H.; Stanciu, L.A. Effects of microstructure and heat treatment on mechanical properties and corrosion behavior of powder metallurgy derived Fe–30Mn alloy. *Mater. Sci. Eng. A* **2017**, *703*, 214–226. [CrossRef]
21. Schinhammer, M.; Pecnik, C.M.; Rechberger, F.; Hänzi, A.C.; Löffler, J.F.; Uggowitzer, P.J. Recrystallization behavior, microstructure evolution and mechanical properties of biodegradable Fe–Mn–C(–Pd) TWIP alloys. *Acta Mater.* **2012**, *60*, 2746–2756. [CrossRef]
22. Niendorf, T.; Brenne, F.; Hoyer, P.; Schwarze, D.; Schaper, M.; Grothe, R.; Wiesener, M.; Grundmeier, G.; Maier, H.J. Processing of New Materials by Additive Manufacturing: Iron-Based Alloys Containing Silver for Biomedical Applications. *Metall. Mater. Trans. A* **2015**, *46*, 2829–2833. [CrossRef]

23. Bouaziz, O.; Allain, S.; Scott, C.P.; Cugy, P.; Barbier, D. High manganese austenitic twinning induced plasticity steels: A review of the microstructure properties relationships. *Curr. Opin. Solid State Mater. Sci.* **2011**, *15*, 141–168. [CrossRef]
24. Schinhammer, M.; Hänzi, A.C.; Löffler, J.F.; Uggowitzer, P.J. Design strategy for biodegradable Fe-based alloys for medical applications. *Acta Biomater.* **2010**, *6*, 1705–1713. [CrossRef] [PubMed]
25. Shuai, C.; Yang, W.; Yang, Y.; Pan, H.; He, C.; Qi, F.; Xie, D.; Liang, H. Selective laser melted Fe-Mn bone scaffold: Microstructure, corrosion behavior and cell response. *Mater. Res. Express* **2019**, *7*, 015404. [CrossRef]
26. Crossgrove, J.; Zheng, W. Manganese toxicity upon overexposure. *NMR Biomed.* **2004**, *17*, 544–553. [CrossRef] [PubMed]
27. Schinhammer, M.; Gerber, I.; Hänzi, A.C.; Uggowitzer, P.J. On the cytocompatibility of biodegradable Fe-based alloys. *Mater. Sci. Eng. C* **2013**, *33*, 782–789. [CrossRef] [PubMed]
28. Drynda, A.; Hassel, T.; Bach, F.W.; Peuster, M. In vitro and in vivo corrosion properties of new iron–manganese alloys designed for cardiovascular applications. *J. Biomed. Mater. Res. Part B Appl. Biomater.* **2014**, *103*, 649–660. [CrossRef]
29. Kraus, T.; Moszner, F.; Fischerauer, S.; Fiedler, M.; Martinelli, E.; Eichler, J.; Witte, F.; Willbold, E.; Schinhammer, M.; Meischel, M.; et al. Biodegradable Fe-based alloys for use in osteosynthesis: Outcome of an in vivo study after 52 weeks. *Acta Biomater.* **2014**, *10*, 3346–3353. [CrossRef]
30. Paul, B.; Lode, A.; Placht, A.M.; Voß, A.; Pilz, S.; Wolff, U.; Oswald, S.; Gebert, A.; Gelinsky, M.; Hufenbach, J. Cell−Material Interactions in Direct Contact Culture of Endothelial Cells on Biodegradable Iron-Based Stents Fabricated by Laser Powder Bed Fusion and Impact of Ion Release. *ACS Appl. Mater. Interfaces* **2021**, *14*, 439–451. [CrossRef]
31. He, J.; He, F.L.; Li, D.W.; Liu, Y.L.; Liu, Y.Y.; Ye, Y.J.; Yin, D.C. Advances in Fe-based biodegradable metallic materials. *RSC Adv.* **2016**, *6*, 112819–112838. [CrossRef]
32. Feng, Q.; Zhang, D.; Xin, C.; Liu, X.; Lin, W.; Zhang, W.; Chen, S.; Sun, K. Characterization and in vivo evaluation of a bio-corrodible nitrided iron stent. *J. Mater. Sci. Mater. Med.* **2013**, *24*, 713–724. [CrossRef]
33. Pierson, D.; Edick, J.; Tauscher, A.; Pokorney, E.; Bowen, P.; Gelbaugh, J.; Stinson, J.; Getty, H.; Lee, C.H.; Drelich, J.; et al. A simplified in vivo approach for evaluating the bioabsorbable behavior of candidate stent materials. *J. Biomed. Mater. Res. Part B Appl. Biomater.* **2011**, *100B*, 58–67. [CrossRef]
34. Waksman, R.; Pakala, R.; Baffour, R.; Seabron, R.; Hellinga, D.; Tio, F.O. Short-Term Effects of Biocorrodible Iron Stents in Porcine Coronary Arteries. *J. Interv. Cardiol.* **2007**, *21*, 15–20. [CrossRef] [PubMed]
35. Dargusch, M.S.; Venezuela, J.; Dehghan-Manshadi, A.; Johnston, S.; Yang, N.; Mardon, K.; Lau, C.; Allavena, R. In Vivo Evaluation of Bioabsorbable Fe-35Mn-1Ag: First Reports on In Vivo Hydrogen Gas Evolution in Fe-Based Implants. *Adv. Healthc. Mater.* **2020**, *10*, 2000667. [CrossRef] [PubMed]
36. Schinhammer, M.; Steiger, P.; Moszner, F.; Löffler, J.F.; Uggowitzer, P.J. Degradation performance of biodegradable Fe\Mn\C(\Pd) alloys. *Mater. Sci. Eng. C* **2013**, *33*, 1882–1893. [CrossRef] [PubMed]
37. Čapek, J.; Msallamová, Š.; Jablonská, E.; Lipovc, J.; Vojtěch, D. A novel high-strength and highly corrosive biodegradable Fe-Pd alloy: Structural, mechanical and in vitro corrosion and cytotoxicity study. *Mater. Sci. Eng. C* **2017**, *79*, 550–562. [CrossRef]
38. Mandal, S.; Kishore, V.; Bose, M.; Nandi, S.K.; Roy, M. In vitro and in vivo degradability, biocompatibility and antimicrobial characteristics of Cu added iron-manganese alloy. *J. Mater. Sci. Technol.* **2021**, *84*, 159–172. [CrossRef]
39. Huang, T.; Cheng, J.; Bian, D.; Zheng, Y. Fe–Au and Fe–Ag composites as candidates for biodegradable stent materials. *J. Biomed. Mater. Res. Part B Appl. Biomater.* **2015**, *104*, 225–240. [CrossRef]
40. Mijnendonckx, K.; Leys, N.; Mahillon, J.; Silver, S.; van Houdt, R. Antimicrobial silver: Uses, toxicity and potential for resistance. *Biometals* **2013**, *26*, 609–621. [CrossRef]
41. Krüger, J.T.; Hoyer, K.P.; Filor, V.; Pramanik, S.; Kietzmann, M.; Meißner, J.; Schaper, M. Novel AgCa and AgCaLa alloys for Fe-based bioresorbable implants with adapted degradation. *J. Alloys Compd.* **2021**, *871*, 159544. [CrossRef]
42. Wiesener, M.; Peters, K.; Taube, A.; Keller, A.; Hoyer, K.-P.; Niendorf, T.; Grundmeier, G. Corrosion properties of bioresorbable FeMn-Ag alloys prepared by selective laser melting. *Mater. Corros.* **2017**, *68*, 1028–1036. [CrossRef]
43. Bagha, P.S.; Khakbiz, M.; Sheibani, S.; Ebrahimi-Barough, S.; Hermawan, H. In Vitro Degradation, Hemocompatibility, and Cytocompatibility of Nanostructured Absorbable Fe−Mn−Ag Alloys for Biomedical Application. *ACS Biomater. Sci. Eng.* **2020**, *6*, 2094–2106. [CrossRef] [PubMed]
44. Wang, C.; Tonna, C.; Mei, D.; Buhagiar, J.; Zheludkevich, M.L.; Lamaka, S.V. Biodegradation behaviour of Fe-based alloys in Hanks' Balanced Salt Solutions: Part II. The evolution of local pH and dissolved oxygen concentration at metal interface. *Bioact. Mater.* **2022**, *7*, 412–425. [CrossRef] [PubMed]
45. Hufenbach, J.; Sander, J.; Kochta, F.; Pilz, S.; Voss, A.; Kühn, U.; Gebert, A. Effect of Selective Laser Melting on Microstructure, Mechanical, and Corrosion Properties of Biodegradable FeMnCS for Implant Applications. *Adv. Eng. Mater.* **2020**, *22*, 2000182. [CrossRef]
46. Heiden, M.; Walker, E.; Stanciu, L. Magnesium, Iron and Zinc Alloys, the Trifecta of Bioresorbable Orthopaedic and Vascular Implantation—A Review. *J. Biotechnol. Biomater.* **2015**, *5*, 1000178. [CrossRef]
47. Hall, D.J.; Pourzal, R.; Jacobs, J.J.; Urban, R.M. Metal wear particles in hematopoietic marrow of the axial skeleton in patients with prior revision for mechanical failure of a hip or knee arthroplasty. *J. Biomed. Mater. Res. Part B Appl. Biomater.* **2019**, *107*, 1930–1936. [CrossRef]
48. Urban, R.M.; Jacobs, J.J.; Tomlinson, M.J.; Gavrilovic, J.; Black, J.; Peoc'h, M. Dissemination of Wear Particles to the Liver, Spleen, and Abdominal Lymph Nodes of Patients with Hip or Knee Replacement. *J. Bone Jt. Surg.* **2000**, *82*, 457–476. [CrossRef]

49. Afifi, M.; Saddick, S.; Zinada, O.A.A. Toxicity of silver nanoparticles on the brain of Oreochromis niloticus and Tilapia zillii. *Saudi J. Biol. Sci.* **2016**, *23*, 754–760. [CrossRef]

50. Sjögren, B.; Lönn, M.; Fremling, K.; Feychting, M.; Nise, G.; Kauppinen, T.; Plato, N.; Wiebert, P.; Gustavsson, P. Occupational exposure to particles and incidence of stroke. *Scand. J. Work Environ. Health* **2013**, *39*, 295–301. [CrossRef]

51. Robertson, S.; Miller, M.R. Ambient air pollution and thrombosis. *Part. Fibre Toxicol.* **2018**, *15*, 1–16. [CrossRef]

52. Andreiev, A.; Hoyer, K.P.; Grydin, O.; Frolov, Y.; Schaper, M. Degradable silver-based alloys. *Mater. Werkst.* **2020**, *51*, 517–530. [CrossRef]

53. Krüger, J.T.; Hoyer, K.P.; Schaper, M. Bioresorbable AgCe and AgCeLa alloys for adapted Fe-based implants. *Mater. Lett.* **2022**, *306*, 130890. [CrossRef]

54. Tortella, G.R.; Rubilar, O.; Durán, N.; Diez, M.C.; Martínez, M.; Parada, J.; Seabra, A.B. Silver nanoparticles: Toxicity in model organisms as an overview of its hazard for human health and the environment. *J. Hazard. Mater.* **2020**, *390*, 121974. [CrossRef]

55. Gameros, L.B.; Chevallier, P.; Sarkissian, A.; Mantovani, D. Silver-based antibacterial strategies for healthcare-associated infections: Processes, challenges, and regulations. An integrated review. *Nanomed. Nanotechnol. Biol. Med.* **2020**, *24*, 102142. [CrossRef] [PubMed]

56. Chiang, W.C.; Tseng, I.; Møller, P.; Hilbert, L.R.; Nielsen, T.T.; Wu, J.K. Influence of silver additions to type 316 stainless steels on bacterial inhibition, mechanical properties, and corrosion resistance. *Mater. Chem. Phys.* **2010**, *119*, 123–130. [CrossRef]

57. Gosheger, G.; Hardes, J.; Ahrens, H.; Streitburger, A.; Buerger, H.; Erren, M.; Gunsel, A.; Kemper, F.H.; Winkelmann, W.; Eiff, C. Silver-coated megaendoprostheses in a rabbit model—An analysis of the infection rate and toxicological side effects. *Biomaterials* **2004**, *25*, 5547–5556. [CrossRef] [PubMed]

58. Savvidou, D.; Kaspiris, A.; Trikoupis, I.; Kakouratos, G.; Goumenos, S.; Melissaridou, D.; Papagelopoulos, P.J. Efficacy of antimicrobial coated orthopaedic implants on the prevention of periprosthetic infections: A systematic review and meta-analysis. *J. Bone Jt. Infect.* **2020**, *5*, 212–222. [CrossRef] [PubMed]

59. Sotoudehbagha, P.; Sheibanib, S.; Khakbiz, M.; Ebrahimi-Barough, S.; Hermawand, H. Novel antibacterial biodegradable Fe-Mn-Ag alloys produced by mechanical alloying. *Mater. Sci. Eng. C* **2018**, *88*, 88–94. [CrossRef]

60. Willbold, E.; Gu, X.; Albert, D.; Kalla, K.; Bobe, K.; Brauneis, M.; Janning, C.; Nellesen, J.; Czayka, W.; Tillmann, W.; et al. Effect of the addition of low rare earth elements (lanthanum, neodymium, cerium) on the biodegradation and biocompatibility of magnesium. *Acta Biomater.* **2015**, *11*, 554–562. [CrossRef]

61. Pagano, G.; Guida, M.; Tommasi, F.; Oral, R. Health effects and toxicity mechanisms of rare earth elements—Knowledge gaps and research prospects. *Ecotoxicol. Environ. Saf.* **2015**, *115*, 40–48. [CrossRef]

62. Quan, J.; Lina, K.; Gua, D. Selective laser melting of silver submicron powder modified 316L stainless steel: Influence of silver addition on microstructures and performances. *Powder Technol.* **2020**, *364*, 478–483. [CrossRef]

63. Krüger, J.T.; Hoyer, K.P.; Hengsbach, F.; Schaper, M. Formation of insoluble silver-phases in an iron-manganese matrix for bioresorbable implants using varying laser-beam-melting-strategies. *J. Mater. Res. Technol.* **2022**, *19*, 2369–2387. [CrossRef]

64. Krüger, J.T. Adjustment of AgCaLa Phases in a FeMn Matrix via LBM for Implants with Adapted Degradation. *Crystals* **2022**, *12*, 1146. [CrossRef]

65. Liu, R.Y.; He, R.G.; Chen, Y.X.; Guo, S.F. Effect of Ag on the Microstructure, Mechanical and Bio-corrosion Properties of Fe–30Mn Alloy. *Acta Metall. Sin.* **2019**, *32*, 1337–1345. [CrossRef]

66. Örnek, C.; Leygraf, C.; Pan, J. On the Volta potential measured by SKPFM—Fundamental and practical aspects with relevance to corrosion science, Corrosion Engineering. *Sci. Technol.* **2019**, *54*, 185–198. [CrossRef]

67. de Faria, D.L.A.; Silva, S.V.; de Oliveira, M.T. Raman microspectroscopy of some iron oxides and oxyhydroxides. *J. Raman Spectrosc.* **1997**, *28*, 873–878. [CrossRef]

68. Testa-Anta, M.; Ramos-Docampo, M.A.; Comesaña-Hermo, M.; Rivas-Murias, B.; Salgueiriño, V. Raman spectroscopy to unravel the magnetic properties of iron oxide nanocrystals for bio-related applications. *Nanoscale Adv.* **2019**, *1*, 2086–2103. [CrossRef]

69. Li, Z.; Huang, J.; Zhong, J.; Li, J. Preparation of AgCl with enhanced photocatalytic activity using ionic liquid as chlorine source. *Appl. Phys. A* **2020**, *126*, 554. [CrossRef]

70. Liebman, J.F. Paradigms and Paradoxes: The Solubility of AgCl in Water: Some Thermochemical Issues of Aqueous Ag+ Ion. *Struct. Chem.* **2004**, *15*, 165–168. [CrossRef]

71. Loffredo, S.; Gambaro, S.; de Andrade, L.M.; Paternoster, C.; Casati, R.; Giguère, N.; Vedani, M.; Mantovani, D. Six-Month Long In Vitro Degradation Tests of Biodegradable Twinning-Induced Plasticity Steels Alloyed with Ag for Stent Applications. *ACS Biomater. Sci. Eng.* **2021**, *7*, 3669–3682. [CrossRef]

72. Pourbaix, M. Atlas of Electrochemical Equilibria. In *Aqueous Solutions*, 2nd ed.; National Association of Corrosion Engineers: Houston, TX, USA, 1974.

73. Saji, V.S.; Lee, C.W. Molybdenum, Molybdenum Oxides, and their Electrochemistry. *ChemSusChem* **2012**, *5*, 1146–1161. [CrossRef]

74. Bagha, P.S.; Khakbiz, M.; Sheibani, S.; Hermawan, H. Design and characterization of nano and bimodal structured biodegradable Fe-Mn-Ag alloy with accelerated corrosion rate. *J. Alloys Compd.* **2018**, *767*, 955–965. [CrossRef]

75. Loffredo, S.; Paternoster, C.; Giguère, N.; Vedani, M.; Mantovani, D. Effect of Silver on Corrosion Behavior of Plastically Deformed Twinning-Induced Plasticity Steel for Biodegradable Stents. *J. Miner. Met. Mater. Soc.* **2020**, *72*, 1892–1901. [CrossRef]

76. Conti, M.C.; Mallia, B.; Sinagra, E.; Wismayer, P.S.; Buhagiar, J.; Vella, D. The effect of alloying elements on the properties of pressed and non-pressed biodegradable Fe–Mn–Ag powder metallurgy alloys. *Heliyon* **2019**, *5*, e02522. [CrossRef]

77. Babacan, N.; Kochta, F.; Hoffmann, V.; Gemming, T.; Kühn, U.; Giebeler, L.; Gebert, A.; Hufenbach, J. Effect of silver additions on the microstructure, mechanical properties and corrosion behavior of biodegradable Fe-30Mn-6Si. *Mater. Commun.* **2021**, *28*, 102689. [CrossRef]
78. Rabeeh, V.P.M.; Hanas, T. Progress in manufacturing and processing of degradable Fe-based implants: A review. *Prog. Biomater.* **2022**, *11*, 163–191. [CrossRef]
79. Huang, T.; Cheng, J.; Zheng, Y.F. In vitro degradation and biocompatibility of Fe–Pd and Fe–Pt composites fabricated by spark plasma sintering. *Mater. Sci. Eng. C* **2014**, *35*, 43–53. [CrossRef]
80. Hermawan, H.; Dubé, D.; Mantovani, D. Development of Degradable Fe-35Mn Alloy for Biomedical Application. *Adv. Mater. Res.* **2007**, *15–17*, 107–112. [CrossRef]
81. Barrett, K.E.; Barman, S.M.; Yuan, J.; Brooks, H.L. *Ganong's Review of Medical Physiology*; McGraw Hill/Medical: New York, NY, USA, 2019.

5 Summary and conclusions

This chapter summarizes the main findings presented in the publications A-F, correlate them to each other, and verifies whether the research hypotheses formulated in chapter 3 are fulfilled. Based on the conclusions of the present work, an outlook for further meaningful research is given.

FeMn needs to be modified with degradable Silver phases. For this purpose, degradable Silver alloys processed via conventional casting are successfully developed (publications A, B). This approach reduces the necessary experimental effort. Based on the properties of promising binary alloys, tertiary alloys are selected for the development of degradable Silver alloys. This strategic development provides an understanding of the impact of alloying elements. Thus, a specific, targeted adaption of the Silver alloy and Silver phases in the FeMn matrix is possible. The developed AgCaLa alloy and AgCeLa alloy exhibit a multiphase structure. This structure is expected to support the dissolution of the alloys due to differences in the electrochemical potential. Both are expected to degrade after the FeMn matrix. The AgCaLa alloy does not contain phases of pure Silver, which is beneficial for the dissolution without the remaining pure Silver. During the immersion tests, the samples continuously lose weight. This indicates that the degradation products are released into the immersion solution, and, probably they can be degraded by the human organism. Accordingly, the Silver alloys are assumed to be bioresorbable. Both alloys provide an electrochemical potential that is nobler than the potential of FeMn and is capable of inducing anodic dissolution. The AgCaLa alloy is selected for further investigations since it contains no pure Silver phases and Calcium is highly biocompatible. The results of publication A confirm the biocompatibility of the AgCaLa alloy. Thus, hypothesis a) (degradable Silver alloy) is successfully validated.

However, processing via LBM affects the Silver alloy, and adaptions to the LBM process are required e. g. due to differences in the solidification conditions. The impact of the LBM process could be significant since the properties of the Silver alloys are highly sensitive e. g. to the amount of the alloying elements. Slight changes impact the electrochemical potential and therefore, the degradation behavior. Thus, keeping the Silver alloy unchanged during the LBM process is a prerequisite for the adaption of the LBM process. The selected alloying elements are highly reactive, which is complicating the production of powder and processing via LBM. The adaption of the Silver powder is an approach to counteract the impacts of the LBM process. A modification of the Silver alloy might also be necessary, when the Silver phases are as intended, but do not evoke the desired effects. However, the formed AgCaLa phases within the Iron and FeMn matrix already reveal satisfying properties, since the higher potential of the AgCaLa phases is confirmed and they dissolve after the FeMn matrix. Thus, the AgCaLa phases inside the FeMn matrix show the desired properties, although the LBM process affects the Silver alloy as alloying elements diffuse between the insoluble components (publications C, D, E). The chemical composition is affected by the loss of Calcium and uptake of Manganese, slag is incorporated into the AgCaLa phases, and the microstructure of the AgCaLa differs from conventionally processed alloys (publication D).

Apart from the properties of the AgCaLa itself, the morphology and distribution of the phases inside the FeMn matrix are decisive. The addition of Silver increases irregularities, nevertheless, regular processes predominantly affected by surface tension result in the formation of homogenously dispersed Silver phases. To summarize the model of the formation of Silver phases: In the first part of the melt pool, the vapor pressure causes the Silver to emerge. Beyond this zone, the Silver agglomerates as the agglomeration reduce the required surface energy. Due to its lower surface tension, Silver is positioned on the FeMn and forms a layer on the top. An outward-directed vortex of the underlying FeMn transports the Silver to the sides of the melt pool. The Silver phases incorporated into the bulk material are parts of the Silver layer remaining during the generation of the next layer. The deep parts of the layer between adjacent melt pools are most likely to remain in the bulk material. The described formation of Silver phases is observed for FeMn modified with Silver (publication C) and is even more pronounced for FeMn modified with AgCaLa (publication D).

The process described before depends on the LBM process parameters as well as on the material itself. The size of Silver phases can be reduced by increasing the hatch distance, which facilitates the inclusion of Silver in the bulk material but increases the porosity. The facilitated inclusion reduces the agglomeration of Silver on the top and the interaction respectively diffusion between the insoluble phases (publications C, D). An identical impact, but less distinctive, is observed for a reduced layer thickness (publication C). However, a reduction of the diffusion of alloying elements is possible, but diffusion cannot be prevented completely. Hence, diffusion must be taken into account for the design of the material. The Silver phases achieved by applying a reduced layer thickness are more compact, which is beneficial for the mechanical properties. In contrast, flat and elongated phases can be generated by the adaption of the direction of gas flow affecting the absorption of laser radiation and the depth of the melt pool (publication C).

Apart from the adaption of the morphology, a graded distribution of the AgCaLa can be obtained by applying certain LBM parameters, causing a selective removal of AgCaLa by spattering. The correlation between spattering, material properties and LBM parameters enables the formation of areas without any AgCaLa (publication D). The material properties influence the LBM process further since the characteristics of the melt pool during LBM significantly differ depending on the material. For pure Iron with Silver and AgCaLa the conduction welding mode (publication E), for FeMn with Silver the transition welding mode (publication C), and for FeMn with AgCaLa the keyhole welding mode (publication D) is the dominating mode. The different welding modes are caused by differences in the vapor pressure of the alloying elements. When the vapor pressure increases, the depression of the melt increases. The highest vapor pressure occurs for FeMn modified with AgCaLa due to the accumulated high vapor pressure of Calcium and Manganese.

Moreover, the material affects the mechanisms responsible for the formation of Silver phases, as a narrow shape of Silver phases is observed for the combination of pure

Iron with pure Silver (publication E). The low difference in surface tension between Iron and Silver favors the inclusion of Silver and reduces the agglomeration of Silver.

In summary, the intended microstructure is achieved for FeMn with AgCaLa processed via LBM since homogenously dispersed Silver phases, almost with the desired chemical composition, are obtained. Furthermore, the mechanical properties are satisfying, as the yield strength is only slightly affected by the modification with AgCaLa and the FeMn matrix remains austenitic, which enables the application of magnetic resonance imaging. Moreover, based on the knowledge obtained, a specific adaptation of the degradable Silver alloy and the LBM process is possible. Thus, hypothesis b) (processability of FeMn with AgCaLa via LBM) is confirmed.

In contrast hypothesis c) (increased degradation by modification with AgCaLa) is neither clearly fulfilled nor disproven. The gaps surrounding the Silver and AgCaLa phases after immersion under neutral pH values and the increased degradation rate due to the addition of AgCaLa under low pH values confirm the proposed anodic dissolution. Nevertheless, the absence of an increased degradation rate under neutral pH values disproves hypothesis c) (publication E, F). The modification of Silver with Calcium and Lanthanum cannot be a reason for this, as the degradation of FeMnAg and FeMnAgCaLa proceeds almost similar. Also, the absence of the proposed potential difference caused by the diffusion of alloying elements is not reasonable, as no acceleration is observed even for pure Iron modified with pure Silver (publication E). The absence of an increased degradation rate is caused by the deposition of degradation products and components of the immersion solution inhibiting further anodic dissolution. Due to these depositions, the Silver and AgCaLa phases cannot fulfill their task. Probably at the initial starting point of the degradation process, the potential difference causes anodic dissolution, and the gaps develop. As the gaps develop, an inhibiting layer is formed and the difference in the electrochemical potential is no longer effective (publication F).

However, the immersion conditions such as the composition of the solution, pH value, amount of solved gases, etc. have a significant impact on the degradation process. This correlation explains the divergence between the expected and observed degradation as well as the contrary results reported in the literature. Accordingly, an increasing effect, not observed for the conditions investigated in this work, might be present for other conditions. Furthermore, the significant dependence on the immersion conditions clarifies the need to develop materials adapted to each application. Therefore, the gained knowledge of the processing of insoluble materials via LBM can be applied to enable a specific adaption.

Anyway, suppression of the layer formation should be addressed and the modification of Silver phases might be a promising approach since less distinctive depositions are observed on AgCaLa than on Silver (publication E). Detailed analysis is required to understand the mechanisms of layer formation and to develop a strategy to suppress them. The current knowledge is not sufficient to define requirements for further adaption of the material. However, once the requirements are defined, knowledge of the relationship between raw material, LBM process, and microstructure will enable a defined adaptation. Moreover, the knowledge can be adapted to other material combinations and applications such as bearings with Lead (Pb) phases. [28] Another possibility is the combination of degradable Silver alloys with a not degradable Iron-based matrix to achieve a release of antibacterial-acting Silver ions from inert implants to prevent implant-related infections. [96][97][102]

In summary hypothesis c) is partially confirmed, as the fundamental strategy is proven, but the impact on the degradation rate is too low. Hypothesis d) is successfully validated as AgCaLa (publication A) and FeMn with AgCaLa (publication F) are supposed to be biocompatible and an antibacterial effect is present for FeMn with AgCaLa (publication F).

Thus, a FeMnAgX alloy containing a degradable Silver alloy is successfully developed, since the material shows the desired properties. However, the inhibiting layers hinder the effectiveness of the AgCaLa phases. Hence, the research hypothesis is not fully accomplished, as the degradation rate is not increased, although the material has the desired properties. Addressing this concern is mandatory to enable the successful application of the developed material. However, there is still the possibility, that the material is suitable for specific applications. But, if adaptions to newly defined requirements are essential, the gained knowledge regarding degradable Silver alloys and the model of the interaction of insoluble phases during LBM respectively the formation of Silver phases during LBM enables an aimed adaption of FeMn modified with a degradable Silver alloy.

6 References

[1] F. Zivic, S. Affatato, M. Trajanovic, M. Schnabelrauch, N. Grujovic, K.L. Choy, Biomaterials in Clinical Practice, Springer, Cham (2018).

[2] D. Carluccio, A.G. Demir, M. J. Bermingham, M. S. Dargusch, Challenges and Opportunities in the Selective Laser Melting of Biodegradable Metals for Load-Bearing Bone Scaffold Applications, Metallurgical and Materials Transactions A 51 (2020) 3311–3334. https://doi.org/10.1007/s11661-020-05796-z

[3] S. Veerachamy, T. Yarlagadda, G. Manivasagam, P. KDV Yarlagadda, Bacterial adherence and biofilm formation on medical implants: A review, Proceedings of the Institution of Mechanical Engineers, Part H: Journal of Engineering in Medicine 228 (2014) 1083–1099. https://doi.org/10.1177/0954411914556137

[4] Q. Chen, G.A. Thouas, Metallic implant biomaterials, Materials Science and Engineering R: Reports 87 (2015) 1-57. http://dx.doi.org/10.1016/j.mser.2014.10.001

[5] B.D. Ratner, A.S. Hoffman, F.J. Schoen, J.E. Lemons, Biomaterials Science, Third ed., Academic Press (2013).

[6] H. Hermawan, Updates on the research and development of absorbable metals for biomedical applications, Progress in Biomaterials 7 (2018) 93–110. https://doi.org/10.1007/s40204-018-0091-4

[7] M. Prakasam, J. Locs, K. Salma-Ancane, D. Loca, A. Largeteau, L. Berzina-Cimdina, Biodegradable Materials and Metallic Implants—A Review, Journal of Functional Biomaterials 8 (2017) 44. https://doi.org/10.3390/jfb8040044

[8] M. Alonzo, F.A. Primo, S.A. Kumar, J.A. Mudloff, E. Dominguez, G. Fregoso, N. Ortiz, W.M. Weiss, B. Joddar, Bone tissue engineering techniques, advances and scaffolds for treatment of bone defects, Current Opinion in Biomedical Engineering 17 (2021) 100248. https://doi.org/10.1016/j.cobme.2020.100248

[9] L. Pleva, P. Kukla, O. Hlinomaz, Treatment of coronary in-stent restenosis: a systematic review, Journal of Geriatric Cardiology 15 (2018) 173-184. https://doi.org/10.11909/j.issn.1671-5411.2018.02.007

[10] Y.F. Zheng, X.N. Gu, F. Witte, Biodegradable metals, Materials Science and Engineering: R: Reports 77 (2014) 1-34. https://doi.org/10.1016/j.mser.2014.01.001

[11] H. Hermawan, D. Dubé, D. Mantovani, Development of Degradable Fe-35Mn Alloy for Biomedical Application, Advanced Materials Research 15-17 (2007) 107-112. http://dx.doi.org/10.4028/www.scientific.net/AMR.15-17.107

[12] N. Abbaspour, R. Hurrell, R. Kelishadi, Review on iron and its importance for human health, Journal of research in medical sciences 19 (2014) 164–174. https://www.ncbi.nlm.nih.gov/pmc/articles/PMC3999603/

[13] M.S. Dargusch, J. Venezuela, A. Dehghan-Manshadi, S. Johnston, N. Yang, K. Mardon, C. Lau, R. Allavena, In Vivo Evaluation of Bioabsorbable Fe-35Mn-1Ag: First Reports on In Vivo Hydrogen Gas Evolution in Fe-Based Implants, Advanced Healthcare Materials 10 (2020). https://doi.org/10.1002/adhm.202000667

[14] S. Mandal, Viraj, S.K. Nandi, M. Roy, Effects of multiscale porosity and pore interconnectivity on in vitro and in vivo degradation and biocompatibility of Fe–Mn–Cu scaffolds, Journal of Materials Chemistry B 9 (2021) 4340-4354. https://doi.org/10.1039/D1TB00641J

[15] W. Lin, L. Qin, H. Qi, D. Zhang, G. Zhang, R. Gao, H. Qiu, Y. Xia, P. Cao, X. Wang, W. Zheng, Long-term in vivo corrosion behavior, biocompatibility and bioresorption mechanism of a bioresorbable nitrided iron scaffold, Acta Biomaterialia 54 (2017) 454-468. http://dx.doi.org/10.1016/j.actbio.2017.03.020

[16] M.S. Dargusch, A. Dehghan-Manshadi, M. Shahbazi, J. Venezuela, X. Tran, J. Song, N. Liu, C. Xu, Q. Ye, C. Wen, Exploring the Role of Manganese on the Microstructure, Mechanical Properties, Biodegradability, and Biocompatibility of Porous Iron-Based Scaffolds, ACS Biomaterials Science & Engineering 5 (2019) 1686–1702. https://doi.org/10.1021/acsbiomaterials.8b01497

[17] M. Dehestani, K. Trumble, H. Wang, H. Wang, L.A. Stanciu, Effects of microstructure and heat treatment on mechanical properties and corrosion behavior of powder metallurgy derived Fe–30Mn alloy, Materials Science & Engineering A 703 (2017) 214-226. http://dx.doi.org/10.1016/j.msea.2017.07.054

[18] C. Shuai, W. Yang, Y. Yang, H. Pan, C. He, F. Qi, D. Xie, H. Liang, Selective laser melted Fe-Mn bone scaffold: microstructure, corrosion behavior and cell response, Materials Research Express 7 (2019) 015404. http://dx.doi.org/10.1088/2053-1591/ab62f5

[19] A. Drynda, T. Hassel, F.W. Bach, M. Peuster, In vitro and in vivo corrosion properties of new iron–manganese alloys designed for cardiovascular applications, Journal of Biomedical Materials Research Part B Applied Biomaterials 103 (2014) 649-660. https://doi.org/10.1002/jbm.b.33234

[20] T. Niendorf, F. Brenne, P. Hoyer, D. Schwarze, M. Schaper, R. Grothe, M. Wiesener, G. Grundmeier, H.J. Maier, Processing of New Materials by Additive Manufacturing: Iron-Based Alloys Containing Silver for Biomedical Applications, Metallurgical and Materials Transactions A 46 (2015) 2829–2833. https://doi.org/10.1007/s11661-015-2932-2

[21] T. Huang, J. Cheng, D. Bian, Y. Zheng, Fe–Au and Fe–Ag composites as candidates for biodegradable stent materials, Journal of Biomedical Materials Research Part B: Applied Biomaterials 104 (2015) 225-240. https://doi.org/10.1002/jbm.b.33389

[22] T. Huang, J. Cheng, Y.F. Zheng, In vitro degradation and biocompatibility of Fe–Pd and Fe–Pt composites fabricated by spark plasma sintering, Materials Science and Engineering C 35 (2014) 43-53. http://dx.doi.org/10.1016/j.msec.2013.10.023

[23] J. Čapek, Š. Msallamová, E. Jablonská, J. Lipovc, D. Vojtěch, A novel high-strength and highly corrosive biodegradable Fe-Pd alloy: Structural, mechanical and in vitro corrosion and cytotoxicity study, Materials Science and Engineering C 79 (2017) 550-562. http://dx.doi.org/10.1016/j.msec.2017.05.100

[24] J. He, F.L. He, D.W. Li, Y.L. Liu, Y.Y. Liu, Y.J. Ye, D.C. Yin, Advances in Fe-based biodegradable metallic materials, RSC Advances 6 (2016) 112819–112838. http://dx.doi.org/10.1039/c6ra20594a

[25] M. Wiesener, K. Peters, A. Taube, A. Keller, K.-P. Hoyer, T. Niendorf, G. Grundmeier, Corrosion properties of bioresorbable FeMn-Ag alloys prepared by selective laser melting, Materials and Corrosion 68 (2017) 1028-1036. https://doi.org/10.1002/maco.201709478

[26] P.S. Bagha, M. Khakbiz, S. Sheibani, H. Hermawan, Design and characterization of nano and bimodal structured biodegradable Fe-Mn-Ag alloy

with accelerated corrosion rate, Journal of Alloys and Compounds 767 (2018) 955-965. https://doi.org/10.1016/j.jallcom.2018.07.206

[27] ASM Alloy P D D ASM Alloy Phase Diagram Database (1999).

[28] J.Z. Zhao, T. Ahmed, H.X. Jiang, J. He, Q. Sun, Solidification of Immiscible Alloys: A Review, Acta Metallurgica Sinica 30 (2017) 1–28. https://doi.org/10.1007/s40195-016-0523-x

[29] C. Wei, J. Wang, Y. He, J. Li, E. Beaugnon, Solidification of Immiscible Alloys under High Magnetic Field: A Review, Metals 11 (2021) 525. https://doi.org/10.3390/met11030525

[30] J. Kraner, J. Medved, M. Godec, I. Paulin, Thermodynamic Behavior of Fe-Mn and Fe-Mn-Ag Powder Mixtures during Selective Laser Melting, Metals 11 (2021) 234. https://doi.org/10.3390/met11020234

[31] J. Quan, K. Lina, D. Gua, Selective laser melting of silver submicron powder modified 316L stainless steel: Influence of silver addition on microstructures and performances, Powder Technology 364 (2020) 478–483. https://doi.org/10.1016/j.powtec.2020.01.082

[32] S. Dadbakhsh, R. Mertens, L. Hao, J. Van Humbeeck, J.P. Kruth, Selective Laser Melting to Manufacture "In Situ" Metal Matrix Composites: A Review, Advanced Engineering Materials 21 (2019) 1801244. https://doi.org/10.1002/adem.201801244

[33] C. Mauffrey, B. Herbert, H. Young, M. L. Wilson, M. Hake, P. F. Stahel, The role of biofilm on orthopaedic implants: the "Holy Grail" of post-traumatic infection management?, European Journal of Trauma and Emergency Surgery 42 (2016) 411–416. https://doi.org/10.1007/s00068-016-0694-1

[34] S. Veerachamy, T. Yarlagadda, G. Manivasagam, P.K. Yarlagadda, Bacterial adherence and biofilm formation on medical implants: A review, Proceedings of the Institution of Mechanical Engineers, Part H: Journal of Engineering in Medicine 228, 10 (2014) 1083-1099. https://doi.org/10.1177/0954411914556137

[35] D.R. Garcia, D.G. Deckey, A. Zega, C. Mayfield, C. S L Spake, T. Emanuel, A. Daniels, J. Jarrell, J. Glasser, C.T. Born, C. P. Eberso, Analysis of growth and biofilm formation of bacterial pathogens on frequently used spinal implant materials, Spine Deformity 8 (2020) 351-359. https://doi.org/10.1007/s43390-020-00054-z

[36] D. Leordean, C. Dudescu, T. Marcu, P. Berce, N. Balc, Customized implants with pecific properties, made by selective laser melting, Rapid Prototyping Journal 21 (2015) 98-104. https://doi.org/10.1108/RPJ-11-2012-0107

[37] F.G. Echeverrigaray, S. Echeverrigaray, A.P.L. Delamare, C.H. Wanke, C.A. Figueroa, I.J.R. Baumvol, C. Aguzzoli, Antibacterial properties obtained by low-energy silver implantation in stainless steel surfaces, Surface & Coatings Technology 3 (2016) 345-351. http://dx.doi.org/10.1016/j.surfcoat.2016.09.005

[38] E. Zhang, X. Zhao, J. Hu, R. Wang, S. Fu, G. Qin, Antibacterial metals and alloys for potential biomedical implants, Bioactive Materials 6 (2021) 2569-2612. https://doi.org/10.1016/j.bioactmat.2021.01.030

[39] C. Redlich, Grundlegende werkstofftechnische Forschungsarbeiten zur Entwicklung bioresorbierbarer metallischer Implantatwerkstoffe, Thelem Universitätsverlag & Buchhandel, Dresden (2018).

[40] A. Colombo, E. Karvouni, Biodegradable Stents "Fulfilling the Mission and Stepping Away", Circulation 102 (2000) 371. https://doi.org/10.1161/01.cir.102.4.371

[41] M. Moravej, D. Mantovani, Biodegradable Metals for Cardiovascular Stent Application: Interests and New Opportunities, International Journal of Molecular Sciences 12 (2011) 4250–4270. https://doi.org/10.3390/ijms12074250

[42] Y. Onuma, P. Serruys, Bioresorbable Scaffolds: From Basic Concept to Clinical Applications, Productivity Press, Portland (2017).

[43] C. Redlich, A. Schauer, J. Scheibler, G. Poehle, P. Barthel, A. Maennel, V. Adams, T. Weissgaerber, A. Linke, P. Quadbeck, In Vitro Degradation Behavior and Biocompatibility of Bioresorbable Molybdenum, Metals 11 (2021) 761. https://doi.org/10.3390/met11050761

[44] E. Willbold, X. Gu, D. Albert, K. Kalla, K. Bobe, M. Brauneis, C. Janning, J. Nellesen, W. Czayka, W. Tillmann, Y. Zheng, F. Witte, Effect of the addition of low rare earth elements (lanthanum, neodymium, cerium) on the biodegradation and biocompatibility of magnesium, Acta Biomaterialia 11 (2015) 554–562, https://doi.org/10.1016/j.actbio.2014.09.041

[45] P. Trumbo, S. Schlicker, A.A. Yates, M. Poos, Dietary reference intakes for energy, carbohydrate, fiber, fat, fatty acids, cholesterol, protein and amino acids, Journal of the American Dietetic Association 102 (2002) 1621-1630. https://doi.org/10.1016/S0002-8223(02)90346-9

[46] G. Poologasundarampillai, A. Nommeots-Nomm, Materials for 3D printing in medicine: metals, polymers, ceramics, hydrogels, 3D Printing in Medicine (2017) 43-71. https://doi.org/10.1016/B978-0-08-100717-4.00002-8

[47] K.H.C. Lim, L.J. Riddell, C.A. Nowson, A.O. Booth, E.A. Szymlek-Gay, Iron and Zinc Nutrition in the Economically-Developed World: A Review, Nutrients 5 (2013) 3184-3211. https://doi.org/10.3390/nu5083184

[48] M. Peuster, C. Hesse, T. Schloo, C. Fink, P. Beerbaum, C. von Schnakenburg, Long-term biocompatibility of a corrodible peripheral iron stent in the porcine descending aorta, Biomaterials 27 (2006) 4955-4962. https://doi.org/10.1016/j.biomaterials.2006.05.029

[49] C. Tonna, C. Wang, D. Mei, S.V. Lamaka, M.L. Zheludkevich, J. Buhagiar, Biodegradation behaviour of Fe-based alloys in Hanks' Balanced Salt Solutions: Part I. material characterisation and corrosion testing, Bioactive Materials 7 (2022) 426-440. https://doi.org/10.1016/j.bioactmat.2021.05.048

[50] H. Hermawan, H. Alamdari, D. Mantovani, D. Dube, Iron–manganese: new class of metallic degradable biomaterials prepared by powder metallurgy, Powder Metallurgy 51 (2008) 1-11. http://dx.doi.org/10.1179/174329008X284868"

[51] E. Houdremont, Einführung in die Sonderstahlkunde, Springer, Berlin (1935).

[52] O. Bouaziz, S. Allain, C.P. Scott, P. Cugy, D. Barbier, High manganese austenitic twinning induced plasticity steels: A review of the microstructure properties relationships, Current Opinion in Solid State and Materials Science 15 (2011) 141–168. https://doi.org/10.1016/j.cossms.2011.04.002

[53] H. Hermawan, A. Purnama, D. Dube, J. Couet, D. Mantovani, Fe–Mn alloys for metallic biodegradable stents: Degradation and cell viability studies, Acta Biomaterialia 6 (2010) 1852-1860. http://dx.doi.org/10.1016/j.actbio.2009.11.025

[54] M. Schinhammer, A.C. Hänzi, J.F. Löffler, P.J. Uggowitzer, Design strategy for biodegradable Fe-based alloys for medical applications, Acta Biomaterialia 6 (2010) 1705-1713. http://dx.doi.org/10.1016/j.actbio.2009.07.039

[55] M. Schinhammer, C.M. Pecnik, F. Rechberger, A.C. Hänzi, J.F. Löffler, P.J. Uggowitzer, Recrystallization behavior, microstructure evolution and

mechanical properties of biodegradable Fe–Mn–C(–Pd) TWIP alloys, Acta Materialia 60 (2012) 2746-2756. http://dx.doi.org/10.1016/j.actamat.2012.01.041

[56] J. Hufenbach, J. Sander, F. Kochta, S. Pilz, A. Voss, U. Kühn, A. Gebert, Effect of Selective Laser Melting on Microstructure, Mechanical, and Corrosion Properties of Biodegradable FeMnCS for Implant Applications, Advanced Engineering Materials 22 (2020) 2000182. https://doi.org/10.1002/adem.202000182

[57] H. Hermawan, D. Mantovani, Process of prototyping coronary stents from biodegradable Fe–Mn alloys, Acta Biomaterialia 9 (2013) 8585-8592. https://doi.org/10.1016/j.actbio.2013.04.027"

[58] B. Paul, A. Lode, A.M. Placht, A. Voß, S. Pilz, U. Wolff, S. Oswald, A. Gebert, M. Gelinsky, J. Hufenbach, Cell–Material Interactions in Direct Contact Culture of Endothelial Cells on Biodegradable Iron-Based Stents Fabricated by Laser Powder Bed Fusion and Impact of Ion Release, ACS Applied Materials & Interfaces 14 (2021) 439-451. https://doi.org/10.1021/acsami.1c21901

[59] J. Crossgrove, W. Zheng, Manganese toxicity upon overexposure, NMR in Biomedicine 17 (2004) 544-553. https://doi.org/10.1002/nbm.931

[60] M. Schinhammer, I. Gerber, A.C. Hänzi, P.J. Uggowitzer, On the cytocompatibility of biodegradable Fe-based alloys, Materials Science and Engineering C 33 (2013) 782-789. http://dx.doi.org/10.1016/j.msec.2012.11.002

[61] J. A Roth, Homeostatic and toxic mechanisms regulating manganese uptake, retention, and elimination, Biological Research 39(1) (2006) 47-57. https://doi.org/10.4067/s0716-97602006000100006

[62] C. Wang, C. Tonna, D. Mei, J. Buhagiar, M.L. Zheludkevich, S.V. Lamaka, Biodegradation behaviour of Fe-based alloys in Hanks' Balanced Salt Solutions: Part II. The evolution of local pH and dissolved oxygen concentration at metal interface, Bioactive Materials 7 (2022) 412–425. https://doi.org/10.1016/j.bioactmat.2021.05.014

[63] D. Pierson, J. Edick, A. Tauscher, E. Pokorney, P. Bowen, J. Gelbaugh, J. Stinson, H. Getty, C.H. Lee, J. Drelich, J. Goldman, A simplified in vivo approach for evaluating the bioabsorbable behavior of candidate stent materials, Journal of Biomedical Materials Research Part B: Applied Biomaterials 100B (2011) 58-67. https://doi.org/10.1002/jbm.b.31922

[64] M. Schinhammer, P. Steiger, F. Moszner, J.F. Löffler, P.J. Uggowitzer, Degradation performance of biodegradable Fe\Mn\C(\Pd) alloys, Materials Science and Engineering C 33 (2013) 1882-1893. http://dx.doi.org/10.1016/j.msec.2012.10.013

[65] P.S. Bagha, M. Khakbiz, S. Sheibani, S. Ebrahimi-Barough, H. Hermawan, In Vitro Degradation, Hemocompatibility, and Cytocompatibility of Nanostructured Absorbable Fe–Mn–Ag Alloys for Biomedical Application, ACS Biomaterials Science and Engineering (2020) 2094–2106. https://dx.doi.org/10.1021/acsbiomaterials.0c00263

[66] T. Kraus, F. Moszner, S. Fischerauer, M. Fiedler, E. Martinelli, J. Eichler, F. Witte, E. Willbold, M. Schinhammer, M. Meischel, P.J. Uggowitzer, J.F. Löffler, A. Weinberg, Biodegradable Fe-based alloys for use in osteosynthesis: Outcome of an in vivo study after 52 weeks, Acta Biomaterialia 10 (2014) 3346-3353. http://dx.doi.org/10.1016/j.actbio.2014.04.007

[67] K.E. Barrett, S.M. Barman, J. Yuan, H.L. Brooks, Ganong`s Review of medical Physiology, McGraw Hill / Medical, New York City (2019).

[68] M. Pourbaix, Atlas of electrochemical equilibria- In aqueous solutions, Second English Edition, National Association of Corrosion Engineers, Houston (1974).

[69] V.S. Saji, C.W. Lee, Molybdenum, Molybdenum Oxides, and their Electrochemistry, ChemSusChem 5 (2012) 1146-1161. https://doi.org/10.1002/cssc.201100660

[70] Q. Feng, D. Zhang, C. Xin, X. Liu, W. Lin, W. Zhang, S. Chen, K. Sun, Characterization and in vivo evaluation of a bio-corrodible nitrided iron stent, Journal of Materials Science: Materials in Medicine 24 (2013) 713–724. https://doi.org/10.1007/s10856-012-4823-z

[71] R. Waksman, R. Pakala, R. Baffour, R. Seabron, D. Hellinga, F.O. Tio, Short-Term Effects of Biocorrodible Iron Stents in Porcine Coronary Arteries, Journal of Interventional Cardiology 21 (2007) 15-20. https://doi.org/10.1111/j.1540-8183.2007.00319.x

[72] D.J. Hall, R. Pourzal, J.J. Jacobs, R.M. Urban, Metal wear particles in hematopoietic marrow of the axial skeleton in patients with prior revision for mechanical failure of a hip or knee arthroplasty, Journal of Biomedical Materials Research Part B Applied Biomaterials 107 (2019) 1930-1936. https://doi.org/10.1002/jbm.b.34285

[73] A.K. Madl, M. Liong, M. Kovochich, B.L. Finley, D.J. Paustenbach, G. Oberdörster, Toxicology of wear particles of cobalt-chromium alloy metal-on-metal hip implants Part I: Physicochemical properties in patient and simulator studies, Nanomedicine: Nanotechnology, Biology and Medicine 11 (2015) 1201-1215. https://doi.org/10.1016/j.nano.2014.12.005

[74] R.M. Urban, J.J. Jacobs, M.J. Tomlinson, J. Gavrilovic, J. Black, M. Peoc'h, Dissemination of Wear Particles to the Liver, Spleen, and Abdominal Lymph Nodes of Patients with Hip or Knee Replacement, The Journal of Bone and Joint Surgery 82 (2000) 457-76. https://doi.org/10.2106/00004623-200004000-00002

[75] M. Afifi, S. Saddick, O. A. Abu Zinada, Toxicity of silver nanoparticles on the brain of Oreochromis niloticus and Tilapia zillii, Saudi Journal of Biological Sciences 23 (2016) 754-760. https://doi.org/10.1016/j.sjbs.2016.06.008

[76] B. Sjögren, M. Lönn, K. Fremling, M. Feychting, G. Nise, T. Kauppinen, N. Plato, P. Wiebert, P. Gustavsson, Occupational exposure to particles and incidence of stroke, Scandinavian Journal of Work, Environment & Health 39 (2013) 295-301. https://doi.org/10.5271/sjweh.3271

[77] S. Robertson, M.R. Miller, Ambient air pollution and thrombosis, Particle and Fibre Toxicology 15 (2018). https://doi.org/10.1186/s12989-017-0237-x

[78] B.D. Ratner, The Biocompatibility of Implant Materials, in: S.F. Badylak, Host Response to Biomaterials, Academic Press (2015) 37-51. https://doi.org/10.1016/B978-0-12-800196-7.00003-7

[79] B.A. Fiedler, Managing Medical Devices Within a Regulatory Framework, Elsevier (2017) 91-108. https://doi.org/10.1016/B978-0-12-804179-6.00006-X.

[80] M. Hellmann, S.D. Mehta, D.M. Bishai, S.C. Mears, J.M. Zenilman, The estimated magnitude and direct hospital costs of prosthetic joint infections in the United States, 1997 to 2004, Journal of Arthroplasty 25 (2009) 766-771. https://doi.org/10.1016/j.arth.2009.05.025

[81] N. Høiby, T Bjarnsholt, M. Givskov, S. Molin, O. Ciofu, Antibiotic resistance of bacterial biofilms, International Journal of Antimicrobial Agents 35 (2010) 322-332. https://doi.org/10.1016/j.ijantimicag.2009.12.011

[82] M.G. Ahlström, J.P. Thyssen, M. Wennervaldt, T. Menné, J.D. Johansen, Nickel allergy and allergic contact dermatitis: A clinical review of immunology,

epidemiology, exposure, and treatment, Contact Dermatitis 81 (2019) 227-241. https://doi.org/10.1111/cod.13327

[83] S. Mandal, V. Kishore, M. Bose, S.K. Nandi, M. Roy, In vitro and in vivo degradability, biocompatibility and antimicrobial characteristics of Cu added iron-manganese alloy, Journal of Materials Science & Technology 84 (2021) 159-172. https://doi.org/10.1016/j.jmst.2020.12.029

[84] K. Mijnendonckx, N. Leys, J. Mahillon, S. Silver, R. Van Houdt, Antimicrobial silver: uses, toxicity and potential for resistance, Biometals 26 (2013) 609–621. https://doi.org/10.1007/s10534-013-9645-z

[85] R.Y. Liu, R.G. He, Y.X. Chen, S.F. Guo, Effect of Ag on the Microstructure, Mechanical and Bio-corrosion Properties of Fe–30Mn Alloy, Acta Metallurgica Sinica 32 (2019) 1337–1345. https://doi.org/10.1007/s40195-019-00911-5

[86] S. Loffredo, C. Paternoster, N. Giguère, M. Vedani, D. Mantovani, Effect of Silver on Corrosion Behavior of Plastically Deformed Twinning-Induced Plasticity Steel for Biodegradable Stents, The Journal of The Minerals, Metals & Materials Society 72 (2020) 1892–1901. https://doi.org/10.1007/s11837-020-04111-w

[87] M. Caligari Conti, B. Mallia, E. Sinagra, P. Schembri Wismayer, J. Buhagiar, D. Vella, The effect of alloying elements on the properties of pressed and non-pressed biodegradable Fe–Mn–Ag powder metallurgy alloys, Heliyon 5 (2019) e02522. https://doi.org/10.1016/j.heliyon.2019.e02522

[88] M. Heiden, E. Walker and L. Stanciu, Magnesium, Iron and Zinc Alloys, the Trifecta of Bioresorbable Orthopaedic and Vascular Implantation - A Review, Journal of Biotechnology & Biomaterials 05 1000178 (2015). http://dx.doi.org/10.4172/2155-952X.1000178

[89] A. Andreiev, K.P. Hoyer, O. Grydin, Y. Frolov, M. Schaper, Degradable silver-based alloys, Materialwissenschaft und Werkstofftechnik 51 (2020) 517-530. https://doi.org/10.1002/mawe.201900191

[90] K.P. Hoyer, M. Schaper, Alloy Design for Biomedical Applications in Additive Manufacturing, In: The Minerals, Metals & Materials Series (eds) TMS 2019 148th Annual Meeting & Exhibition Supplemental Proceedings. The Minerals, Metals & Materials Series. Springer, Cham (2019). https://doi.org/10.1007/978-3-030-05861-6_44

[91] G.R. Tortella, O. Rubilar, N. Durán, M.C. Diez, M. Martínez, J. Parada, A.B. Seabra, Silver nanoparticles: Toxicity in model organisms as an overview of its hazard for human health and the environment, Journal of Hazardous Materials 390 (2020) 121974. https://doi.org/10.1016/j.jhazmat.2019.121974

[92] L. Bonilla Gameros, P. Chevallier, A. Sarkissian, D. Mantovani, Silver-based antibacterial strategies for healthcare-associated infections: Processes, challenges, and regulations. An integrated review, Nanomedicine: Nanotechnology, Biology and Medicine 24 (2020) 102142. https://doi.org/10.1016/j.nano.2019.102142

[93] A. Wadhera, M. Fung, Systemic argyria associated with ingestion of colloidal silver, Dermatology Online Journal 11 (1) (2005). https://doi.org/10.5070/D30832G6D3

[94] A.B.G. Lansdown, A. Williams, How safe is silver in wound care? Journal of Wound Care 13 (2004) 131–136. https://doi.org/10.12968/jowc.2004.13.4.26596

[95] World Health Organization, Silver in drinking-water background document for development of WHO guidelines for drinking-water quality guidelines for

drinking-water quality, second ed., Health Criteria and other Supporting Information 2 World Health Organization, Geneva (1996).

[96] W.C. Chiang, I-S. Tseng, P Møller, L. Rischel Hilbert, T. Tolker Nielsen, J.K. Wu, Influence of silver additions to type 316 stainless steels on bacterial inhibition, mechanical properties, and corrosion resistance, Materials Chemistry and Physics 119 (2010) 123-130. https://doi.org/10.1016/j.matchemphys.2009.08.035

[97] G. Gosheger, J. Hardes, H. Ahrens, A. Streitburger, H. Buerger, M. Erren, A. Gunsel, F.H. Kemper, W. Winkelmann, C. Eiff, Silver-coated megaendoprostheses in a rabbit model—an analysis of the infection rate and toxicological side effects, Biomaterials 25 (2004) 5547-5556. https://doi.org/10.1016/j.biomaterials.2004.01.008

[98] O. D. Savvidou, A. Kaspiris, I.Trikoupis, G. Kakouratos, S. Goumenos, D. Melissaridou, P. J. Papagelopoulos , Efficacy of antimicrobial coated orthopaedic implants on the prevention of periprosthetic infections: a systematic review and meta-analysis, Journal of Bone and Joint Infection 5 (2020) 212-222. https://doi.org/10.7150/jbji.44839

[99] European Food Safety Authority, Overview on Tolerable Upper Intake Levels as derived by the Scientific Committee on Food (SCF) and the EFSA Panel on Dietetic Products, Nutrition and Allergies (NDA), Summary of Tolerable Upper Intake Levels – version 4 (2018). ⟨https://www.efsa.europa.eu/sites/default/files/assets /UL_Summary_tables.pdf⟩ (Accessed 29.09.2020).

[100] G. Pagano, M. Guida, F. Tommasi, R. Oral, Health effects and toxicity mechanisms of rare earth elements — Knowledge gaps and research prospects, Ecotoxicol. Environ. Saf. 115 (2015) 40–48. https://doi.org/10.1016/j.ecoenv.2015.01.030

[101] S. Hirano, K.T. Suzuki, Exposure, metabolism, and toxicity of rare earths and related compounds, Environ. Health Perspect. 104 (1996) 85–95. https://doi.org/10.1289/ehp.96104s185

[102] P. Sotoudehbagha, S. Sheibanib, M. Khakbiz, S. Ebrahimi-Barough, H. Hermawand, Novel antibacterial biodegradable Fe-Mn-Ag alloys produced by mechanical alloying, Materials Science & Engineering C 88 (2018) 88–94. https://doi.org/10.1016/j.msec.2018.03.005

[103] M. Vignesh, G.R. Kumar, M. Sathishkumar, M. Manikandan, G. Rajyalakshmi, R. Ramanujam, N. Arivazhagan, Development of Biomedical Implants through Additive Manufacturing: A Review, Journal of Materials Engineering and Performance 30 (2021) 4735–4744. https://doi.org/10.1007/s11665-021-05578-7

[104] P. Kumar, D.K. Rajak, M. Abubakar, S.G.M. Ali, M. Hussain, 3D Printing Technology for Biomedical Practice: A Review, Journal of Materials Engineering and Performance 30 (2021) 5342-5355. https://doi.org/10.1007/s11665-021-05792-3

[105] Z. Ferdous, A. Nemmar, Health Impact of Silver Nanoparticles: A Review of the Biodistribution and Toxicity Following Various Routes of Exposure, International Journal of Molecular Sciences 21 (2020). https://doi.org/10.3390/ijms21072375

[106] A. Rodriguez-Contreras, M. Punset, J.A. Calero, F.J. Gil, E. Ruperez, J.M. Manero, Powder metallurgy with space holder for porous titanium implants: A review, Journal of Materials Science & Technology 76 (2021) 129-149. https://doi.org/10.1016/j.jmst.2020.11.005

[107] A.H. Yusop, A.A. Bakir, N.A. Shaharom, M.R. Abdul Kadir, H. Hermawan, Porous Biodegradable Metals for Hard Tissue Scaffolds: A Review, International Journal of Biomaterials 2012 (2012) 641430. https://doi.org/10.1155/2012/641430

[108] S.A. Khairallah, A.T. Anderson, A. Rubenchik, W.E. King, Laser powder-bed fusion additive manufacturing: Physics of complex melt flow and formation mechanisms of pores, spatter, and denudation zones, Acta Materialia 108 (2016) 36-45. http://dx.doi.org/10.1016/j.actamat.2016.02.014

[109] X. Zhou, K.Li, D. Zhang, X. Liu, J. Ma, W. Liu, Z. Shen, Textures formed in a CoCrMo alloy by selective laser melting, Journal of Alloys and Compounds 631 (2015) 153–164. http://dx.doi.org/10.1016/j.jallcom.2015.01.096

[110] S.A. Khairallah, A. Anderson, Mesoscopic simulation model of selective laser melting of stainless steel powder, Journal of Materials Processing Technology 214 (2014) 2627-2636. http://dx.doi.org/10.1016/j.jmatprotec.2014.06.001

[111] E.L. Papazoglou, N. E. Karkalos, P. Karmiris-Obratański, A.P. Markopoulos, On the Modeling and Simulation of SLM and SLS for Metal and Polymer Powders: A Review, Archives of Computational Methods in Engineering 29 (2022) 941–973. https://doi.org/10.1007/s11831-021-09601-x

[112] G. Zhang, J. Chen, M. Zheng, Z. Yan, X. Lu, X. Lin, W. Huang, Element Vaporization of Ti-6Al-4V Alloy during Selective Laser Melting, Metals 10(4) (2020) 435. https://doi.org/10.3390/met10040435

[113] D. Carluccio, A.G. Demirm L. Caprio, B. Previtali, M.J. Bermingham, M.S. Dargusch, The influence of laser processing parameters on the densification and surface morphology of pure Fe and Fe-35Mn scaffolds produced by selective laser melting, Journal of Manufacturing Processes 40 (2019) 113–121. https://doi.org/10.1016/j.jmapro.2019.03.018

[114] J.H.K. Tan, S.L. Sing, W.Y. Yeong, Microstructure modelling for metallic additive manufacturing: a review, Virtual and Physical Prototyping 15 (2019) 87-105. https://doi.org/10.1080/17452759.2019.1677345

[115] D. Wang, S. Wu, F. Fu, S. Mai, Y. Yang, Y. Liu, C. Song, Mechanisms and characteristics of spatter generation in SLM processing and its effect on the properties, Materials and Design 117 (2017) 121-130. https://doi.org/10.1016/j.matdes.2017.09.058

[116] C. Qiu, C. Panwisawas, M. Ward, H.C. Basoalto, J.W. Brooks, M.M. Attallah, On the role of melt flow into the surface structure and porosity development during selective laser melting, Acta Materialia 96 (2015) 72-79. http://dx.doi.org/10.1016/j.actamat.2015.06.004

[117] E. Louvis, P. Fox, C.J. Sutcliffe, Selective laser melting of aluminium components, Journal of Materials Processing Technology 211 (2011) 275–284. https://doi.org/10.1016/j.jmatprotec.2010.09.019

[118] H. Gong, H. Gu, K. Zeng, J.J.S. Dilip, D. Pal, B. Stucker, D. Christiansen, J. Beuth, J. Lewandowski, Melt Pool Characterization for Selective Laser Melting of Ti-6Al-4V Pre-alloyed Powder, Solid Freeform Fabrication Symposium (2014) 256-267.

[119] T. Furumoto, K. Egashira, K. Munekage, S. Abe, Experimental investigation of melt pool behaviour during selective laser melting by high speed imaging, CIRP Annals - Manufacturing Technology 67 (2018) 253-256. https://doi.org/10.1016/j.cirp.2018.04.097

[120] M.J. Matthews, G. Guss, S.A. Khairallah, A.M. Rubenchik, P.J. Depond, W. E. King, Denudation of metal powder layers in laser powder bed fusion

processes, Acta Materialia 114 (2016) 33-42.
http://dx.doi.org/10.1016/j.actamat.2016.05.017

[121] V. Gunenthiram, P. Peyre, M. Schneider, M. Dal, F. Coste, I. Koutiri, R. Fabbro, Experimental analysis of spatter generation and melt-pool behavior during the powder bed laser beam melting process, Journal of Materials Processing Technology 251 (2017) 376-386. http://dx.doi.org/10.1016/j.jmatprotec.2017.08.012

[122] I. Yadroitsev, P. Krakhmalev, I. Yadroitsava, S. Johansson, I. Smurov, Energy input effect on morphology and microstructure of selective laser melting single track from metallic powder, Journal of Materials Processing Technology 213 (2013) 606-613. http://dx.doi.org/10.1016/j.jmatprotec.2012.11.014

[123] T. Heeling, M. Cloots, K. Wegener, Melt Pool Simulation for the Evaluation of Process Parameters in Selective Laser Melting, Additive Manufacturing 14 (2017) 116-125. https://doi.org/10.1016/j.addma.2017.02.003

[124] V. Gunenthiram, P. Peyre, M. Schneider, M. Dal, F. Coste, R. Fabbro, Analysis of laser–melt pool–powder bed interaction during the selective laser melting of a stainless steel, Journal of Laser Applications 29 (2017) 022303. http://dx.doi.org/10.2351/1.4983259

[125] O. Andreau, I. Koutiri, P. Peyre, J.D. Penot, N. Saintier, E. Pessard, T. De Terris, C. Dupuy, T. Baudin, Texture control of 316L parts by modulation of the melt pool morphology in selective laser melting, Journal of Materials Processing Technology 264 (2019) 21-31. https://doi.org/10.1016/j.jmatprotec.2018.08.049

[126] A. Bin Anwar, Q.C. Pham, Selective laser melting of AlSi10Mg: Effects of scan direction, part placement and inert gas flow velocity on tensile strength, Journal of Materials Processing Technology 240 (2017) 388-396. http://dx.doi.org/10.1016/j.jmatprotec.2016.10.015

[127] C. Panwisawas, C. Qiu, M.J. Anderson, Y. Sovani, R.P. Turner, M.M. Attallah, J.W. Brooks, H.C. Basoalto, Mesoscale modelling of selective laser melting: Thermal fluid dynamics and microstructural evolution, Computational Materials Science 126 (2017) 479-490. http://dx.doi.org/10.1016/j.commatsci.2016.10.011

[128] Y.C. Wu, C.H. San, C.H. Chang, H.J. Lin, R. Marwan, S. Baba, W.S. Hwang, Numerical modeling of melt-pool behavior in selective laser melting with random powder distribution and experimental validation, Journal of Materials Processing Technology 254 (2018) 72-78. https://doi.org/10.1016/j.jmatprotec.2017.11.032

[129] S. Patel, M. Vlasea, Melting modes in laser powder bed fusion, Materialia 9 (2020) 100591. https://doi.org/10.1016/j.mtla.2020.100591

[130] W.E. King, H.D. Barth, V.M. Castillo, G.F. Gallegos, J.W. Gibbs, D.E. Hahn, C. Kamath, A.M. Rubenchik, Observation of keyhole-mode laser melting in laser powder-bed-fusion additive manufacturing, Journal of Materials Processing Technology 214 (2014) 2915-2925. http://dx.doi.org/10.1016/j.jmatprotec.2014.06.005

[131] S. Shrestha, Y.K. Chou, by A Numerical ASME Study on the Keyhole Formation During Laser Powder Bed Fusion Process, Journal of Manufacturing Science and Engineering 141(10) (2019) 101002. https://doi.org/10.1115/1.4044100

[132] L. Huanga, X. Hua, D. Wu, L. Fang, Y. Cai, Y. Ye, Effect of magnesium content on keyhole-induced porosity formation and distribution in aluminum

alloys laser welding, Journal of Manufacturing Processes 33 (2018) 43–53, https://doi.org/10.1016/j.jmapro.2018.04.023

[133] A. Queva, G. Guillemot, C. Moriconi, C. Metton, M. Bellet, Numerical study of the impact of vaporisation on melt pool dynamics in Laser Powder Bed Fusion - Application to IN718 and Ti–6Al–4V, Additive Manufacturing 35 (2020) 101249. https://doi.org/10.1016/j.addma.2020.101249

[134] J.H. Cho, S.J. Na, Implementation of real-time multiple reflection and Fresnel absorption of laser beam in keyhole, Journal of Physics D: Applied Physics 39 (2006) 5372–5378. https://doi.org/10.1088/0022-3727/39/24/039

[135] M. Miyagi, H. Wang, R. Yoshida, Y. Kawahito, H. Kawakami, T. Shoubu, Effect of alloy element on weld pool dynamics in laser welding of aluminum alloys, Scientific Reports 8 (2018) 12944. https://doi.org/10.1038/s41598-018-31350-4

[136] K.Q. Le, C. Tang, C.H. Wong, On the study of keyhole-mode melting in selective laser melting process, International Journal of Thermal Sciences 145 (2019) 105992. https://doi.org/10.1016/j.ijthermalsci.2019.105992

[137] A.U. Rehman, M.A. Mahmood, F. Pitir, M. U. Salamci, A.C. Popescu, I.N. Mihailescu, Keyhole Formation by Laser Drilling in Laser Powder Bed Fusion of Ti6Al4V Biomedical Alloy: Mesoscopic Computational Fluid Dynamics Simulation versus Mathematical Modelling Using Empirical Validation, Nanomaterials 11 (2021) 3284. https://doi.org/10.3390/nano11123284

[138] J. Trapp, A.M. Rubenchik, G. Guss, M.J. Matthews, In situ absorptivity measurements of metallic powders during laser powder-bed-fusion additive manufacturing, Applied Materials Today 9 (2017) 341–349. http://dx.doi.org/10.1016/j.apmt.2017.08.006

[139] T. Hibiya, S. Ozawa, Effect of oxygen partial pressure on the marangoni flow of molten metals, Crystal Research and Technology 48 (2013) 208-213. https://doi.org/10.1002/crat.201200514

[140] I. Egry, E. Ricci, R. Novakovic, S. Ozawa, Surface tension of liquid metals and alloys — Recent developments, Advances in Colloid and Interface Science 159 (2010) 198–212. https://doi.org/10.1016/j.cis.2010.06.009

[141] D.W. Cho, Y.D. Park, M. Cheepu, Numerical simulation of slag movement from Marangoni flow for GMAW with computational fluid dynamics, International Communications in Heat and Mass Transfer 125 (2021) 105243. https://doi.org/10.1016/j.icheatmasstransfer.2021.105243

[142] R.N. Murarka, W.K. Lu, A.E. Hamielec, Effect of dissolved oxygen on the surface tension of liquid iron, Canadian Metallurgical Quarterly 14 (1975) 111-115. http://dx.doi.org/10.1179/000844375795050337

[143] H. Yu, S. Hayashi, K. Kakehi, Y.L. Kuo, Study of Formed Oxides in IN718 Alloy during the Fabrication by Selective Laser Melting and Electron Beam Melting, Metals 9 (2019). https://doi.org/10.3390/met9010019

[144] M. Salehi, S. Maleksaeedi, H. Farnoush, N.M.L. Sharon, G.K. Meenashisundaram, M. Gupta, An investigation into interaction between magnesium powder and Ar gas: Implications for selective laser melting of magnesium, Powder Technology 333 (2018) 252-261. https://doi.org/10.1016/j.powtec.2018.04.026

[145] A. Ghasemi, E. Fereiduni, M. Balbaa, S.D. Jadhav, M. Elbestawi , S. Habibi, Influence of alloying elements on laser powder bed fusion processability of aluminum: A new insight into the oxidation tendency, Additive Manufacturing 46 (2021) 102145. https://doi.org/10.1016/j.addma.2021.102145

[146] W. Zhang, L. Wang, Z. Feng, Y. Chen, Research progress on selective laser melting (SLM) of magnesium alloys: A review, Optik 207 (2020) 163842. https://doi.org/10.1016/j.ijleo.2019.163842

[147] L. Cao, S. Chen, M.i Wei, Q. Guo, J. Liang, C. Liu, M. Wang, Effect of laser energy density on defects behavior of direct laser depositing 24CrNiMo alloy steel, Optics and Laser Technology 111 (2019) 541–553. https://doi.org/10.1016/j.optlastec.2018.10.025

[148] X. Shi, C. Yan, W. Feng, Y. Zhang, Z. Leng, Effect of high layer thickness on surface quality and defect behavior of Ti-6Al-4V fabricated by selective laser melting, Optics and Laser Technology 132 (2020) 106471. https://doi.org/10.1016/j.optlastec.2020.106471

[149] M. Lutter-Günther, M. Bröker, T. Mayer, S. Lizak, C. Seidel, G. Reinhart, Spatter formation during laser beam melting of AlSi10Mg and effects on powder quality, Procedia CIRP 74 (2018) 33-38. https://doi.org/10.1016/j.procir.2018.08.008

[150] M.T. Andani, R. Dehghani, M.R. Karamooz-Ravari, R. Mirzaeifar, Jun Ni, A study on the effect of energy input on spatter particles creationduring selective laser melting process, Additive Manufacturing 20 (2018) 33–43. https://doi.org/10.1016/j.addma.2017.12.009

[151] C. Panwisawas, Y. Sovani, R. P. Turner, J. W. Brooks, H. C. Basoalto, I. Choquet, Modelling of thermal fluid dynamics for fusion welding, Journal of Materials Processing Technology 252 (2018) 176–182. http://dx.doi.org/10.1016/j.jmatprotec.2017.09.019

[152] H. Chen, W. Yan, Spattering and denudation in laser powder bed fusion process: Multiphase flow modelling, Acta Materialia 196 (2020) 154-167. https://doi.org/10.1016/j.actamat.2020.06.033

[153] J. Volpp, Spattering effects during selective laser melting, Journal of Laser Applications 32 (2020) 022023. https://doi.org/10.2351/7.0000061

[154] X. Guo, C. Zhang, Q. Tian, D. Yu, Liquid metals dealloying as a general approach for the selective extraction of metals and the fabrication of nanoporous metals: A review, Materials Today Communications 26 (2021) 102007. https://doi.org/10.1016/j.mtcomm.2020.102007

[155] I. McCue, B. Gaskey, P.A. Geslin, A. Karma, J. Erlebacher, Kinetics and morphological evolution of liquid metal dealloying, Acta Materialia 115 (2016) 10-23. http://dx.doi.org/10.1016/j.actamat.2016.05.032

[156] G. Mikhailov, L. Makrovets, Thermodynamic analysis of steel deoxidation with calcium and aluminum, Russian Metallurgy (2008) 727-72. https://doi.org/10.1134/S003602950808017X

[157] K. Taguchi, H. Ono Nakazato, D. Nakai, T. Usui, K. Marukawa, Deoxidation and Desulfurization Equilibria of Liquid Iron by Calcium, ISIJ Internationa 43 (2003) 1705–1709. https://doi.org/10.2355/isijinternational.43.1705

[158] R.J. Fruehan, The effect of zirconium, cerium, and lanthanum on the solubility of oxygen in liquid iron, Metallurgical Transactions 5 (1974) 345-347. https://doi.org/10.1007/BF02644100

[159] W.M. Garrison Jr., J.L. Maloney, Lanthanum additions and the toughness of ultra-high strength steels and the determination of appropriate lanthanum additions, Materials Science and Engineering A 403 (2005) 299–310. https://doi.org/10.1016/j.msea.2005.05.021

[160] B.I. Imasogie, A.A. Afonja, J.A. Ali, Properties of ductile cast iron nodularised with multiple calcium–magnesium based master alloy, Materials Science and Technology 16 (2000) 194-201. https://doi.org/10.1179/026708300101507541

[161] N.B. Sing, A. Mostavan, E. Hamzah, D. Mantovani, H. Hermawan, Degradation behavior of biodegradable Fe35Mn alloy stents, Journal of Biomedical Materials Research Part B: Applied Biomaterial 103 (2014) 572-577. https://doi.org/10.1002/jbm.b.33242

[162] E. Scarcello, D. Lison, Are Fe-Based Stenting Materials Biocompatible? A Critical Review of In Vitro and In Vivo Studies, Journal of Functional Biomaterials 11(2) (2020). https://doi.org/10.3390/jfb11010002

Student theses

The following student work was carried out at the Chair of Materials Science of Paderborn University as part of this dissertation. The definition of the objectives, the processing as well as the evaluation, interpretation, and visualization of results have been carried out under the scientific guidance of the supervisors Prof. Dr.-Ing. habil. Mirko Schaper, Dr.-Ing. Kay-Peter Hoyer and Jan Tobias Krüger. The results obtained have been partly incorporated into the publications included in this dissertation.

[SB19]   S. Beumer, (supervisor: M. Schaper, K.-P. Hoyer): Silberlegierungen zur gesteigerten Degradation bioresorbierbarer Implantatwerkstoffe. Paderborn University, Chair of Materials Science, unpublished bachelor's, thesis, 2019.

[SB20]   S. Beumer, (supervisor: M. Schaper, J.T. Krüger): Bioresorbierbare AgCaLa-Legierungen zur angepassten Degradation eisenbasierter Implantate. Paderborn University, Chair of Materials Science, unpublished student's, thesis, 2020.

[DN20]   D. Niggemann, (supervisor: M. Schaper, J.T. Krüger): Angepasst degradierende AgMgLa-Legierungen für bioresorbierbare Implantate. Paderborn University, Chair of Materials Science, unpublished student's, thesis, 2020.

[KG20]   K. Gossen, (supervisor: M. Schaper, J.T. Krüger): Additiv verarbeitete FeMnAg-Werkstoffe mit angepasstem Degradationsprofil. Paderborn University, Chair of Materials Science, unpublished student's, thesis, 2020.

[EP20]   E.K. Popp, (supervisor: M. Schaper, K.-P. Hoyer, J.T. Krüger): Korrelation von Mikrostruktur und Materialeigenschaften additiv verarbeiteter Fe-Ag-Legierungen für Implantate. Paderborn University, Chair of Materials Science, unpublished master's, thesis, 2020.

[MD20]   M. Dreyer, (supervisor: M. Schaper, K.-P. Hoyer, J.T. Krüger): Degradation additiv verarbeiteter Eisen-Silber-Werkstoffe für resorbierbare Implantate. Paderborn University, Chair of Materials Science, unpublished master's, thesis, 2020.

[SB21]   S. Beumer, (supervisor: M. Schaper, K.-P. Hoyer, J.T. Krüger): Mikrostruktur und Materialeigenschaften additiv verarbeiteter FeMn+AgCaLa-Werkstoffe für resorbierbare Implantate. Paderborn University, Chair of Materials Science, unpublished master's, thesis, 2021.

[JH20]   J. Heinrich, (supervisor: M. Schaper, K.-P. Hoyer, J.T. Krüger): Mikrostruktur und Degradation additiv verarbeiteter Fe+Ag-Werkstoffe für resorbierbare

Implantate. Paderborn University, Chair of Materials Science, unpublished bachelor's, thesis, 2020.

[DN22] D. Niggemann, (supervisor: M. Schaper, J.T. Krüger): Mikrostruktur und Materialeigenschaften additiv verarbei-teter Fe- und Fe+FeMn-Werkstoffe für resorbierbare Implantate. Paderborn University, Chair of Materials Science, unpublished master's, thesis, 2022.

[ME22] M. Ernst, (supervisor: M. Schaper, J.T. Krüger): Etablierung eines Immersionsversuchs zur Analyse der Silberfreisetzung von antibakteriell wirksamen Implantatwerkstoffen. Paderborn University, Chair of Materials Science, unpublished master's, thesis, 2020.

Erklärung zur Zitation von Inhalten aus studentischen Arbeiten

In Ergänzung zu meinem Antrag auf Zulassung zur Promotion in der Fakultät für Maschinenbau der Universität Paderborn erkläre ich gemäß §11 der Promotionsordnung und unter Beachtung der Regelung zur Zitation studentischer Arbeiten:

Die von mir vorgelegte Dissertation habe ich selbstständig verfasst, und ich habe keine anderen als die dort angegebenen Quellen und Hilfsmittel benutzt. Es sind Inhalte studentischen Ursprungs (studentische Arbeiten) in dieser Dissertation enthalten.

Inhalte studentischen Ursprungs sind in die in dieser Dissertation enthaltenen Veröffentlichungen eingeflossen. Die Studierenden habe ich in den Danksagungen der Veröffentlichungen genannt. Darüber hinaus habe ich die verwendeten Arbeiten entsprechend der Regelung „Zitation aus studentischen Arbeiten in Dissertationen" als Quellen aufgeführt.

Paderborn, 12.04.2023   *Jan Tobias Krüger*

Lebenslauf

Persönliche Daten

Name	Jan Tobias Krüger
Geburtsdatum /-ort	26.10.1994, Bielefeld
Staatsangehörigkeit	Deutsch

Beruflicher Werdegang

seit 10/2019	**Wissenschaftlicher Mitarbeiter** Lehrstuhl für Werkstoffkunde (LWK) Universität Paderborn, 33098 Paderborn
05/2019 – 09/2019	**Technischer Mitarbeiter** Lehrstuhl für Werkstoffkunde (LWK) Universität Paderborn, 33098 Paderborn
09/2017 – 05/2019	**Studentische Hilfskraft** Lehrstuhl für Werkstoffkunde (LWK) Universität Paderborn, 33098 Paderborn

Ausbildung

04/2017 – 09/2019	**Masterstudium Maschinenbau** Universität Paderborn, 33098 Paderborn
10/2013 – 03/2017	**Bachelorstudium Maschinenbau** Universität Paderborn, 33098 Paderborn
08/2005 – 07/2013	**Abitur** Marienschule der Ursulinen, 33611 Bielefeld

Auszeichnungen

02/2020	**Preis der Fakultät 2019 für einen herausragenden Abschluss** (Fakultät für Maschinenbau, Universität Paderborn, 33098 Paderborn)
05/2022	**Faculty Best Paper Award 2021** (Fakultät für Maschinenbau, Universität Paderborn, 33098 Paderborn)

Praktika

06/2011	**Sozialpraktikum** Matthias Claudius Kindergarten, 33739 Bielefeld
03/2012	**Verwaltung und Montage** DeKa Kunststofffenster, 33739 Bielefeld
10/2015 – 12/2015	**Grundpraktikum Maschinenbau** Boge Kompressoren, 33739 Bielefeld

Paderborn, 12.04.2023   *Jan Tobias Krüger*